Post-Transition Metals

Edited by Mohammed Muzibur Rahman, Abdullah Mohammed Asiri, Anish Khan, Inamuddin and Thamer Tabbakh

Published in London, United Kingdom

IntechOpen

Supporting open minds since 2005

Post-Transition Metals
http://dx.doi.org/10.5772/intechopen.90968
Edited by Mohammed Muzibur Rahman, Abdullah Mohammed Asiri, Anish Khan, Inamuddin and Thamer Tabbakh

Contributors
Masafumi Yamaguchi, Janmejaya Pradhan, Satya Ranjan Pattanaik, Sunirmal Jana, Hasmat Khan, Saswati Sarkar, Moumita Pal, Susanta Bera, Rabia Benabderrahmane Zaghouani, Ilya Anatolievich Svirsky, Boris Konstantinovich Radionov, Kaleem Ahmad Najar, Saquib Rouf, Sobura Altaf, Shezan Malik, A.M. M Shah, Neerish Revaprasadu, Siphamandla Cecil Masikane, Raghad M. Aljohany, Thamer A. Tabbakh, Hatem Alhazmi, Rawan M. Alsulami, Thierry Baron, Mickael Martin, Yann Bogumilowicz, Deng Huiwen, Keshuang Li, Mingchu Tang, Huiyun Liu

First published in London, United Kingdom, 2021 by IntechOpen
IntechOpen is the global imprint of INTECHOPEN LIMITED, registered in England and Wales, registration number: 11086078, 5 Princes Gate Court, London, SW7 2QJ, United Kingdom
Printed in Croatia

British Library Cataloguing-in-Publication Data
A catalogue record for this book is available from the British Library

Additional hard and PDF copies can be obtained from orders@intechopen.com

Post-Transition Metals
Edited by Mohammed Muzibur Rahman, Abdullah Mohammed Asiri, Anish Khan, Inamuddin and Thamer Tabbakh
p. cm.
Print ISBN 978-1-83968-260-5
Online ISBN 978-1-83968-261-2
eBook (PDF) ISBN 978-1-83968-265-0

For EU product safety concerns:
IN TECH d.o.o., Prolaz Marije Krucifikse Kozulić 3, 51000 Rijeka, Croatia, info@intechopen.com or visit our website at intechopen.com.

Meet the editors

Dr. Mohammed Muzibur Rahman received his BSc and MSc from Shahjalal University of Science and Technology, Sylhet, Bangladesh, in 1999 and 2001, respectively. He received his Ph.D. from Chonbuk National University, South Korea, in 2007. After obtaining his Ph.D., Dr. Rahman worked as a postdoctoral fellow and assistant professor in pioneer research centers and universities in South Korea, Japan, and Saudi Arabia (2007–2011). Presently, he is a full professor at the Center of Excellence for Advanced Materials Research (CEAMR) and the Chemistry Department at King Abdulaziz University, Jeddah, Saudi Arabia. He has published many international and domestic articles and book chapters as well as edited several books. His research interests include transition metals, photocatalysis, electrocatalysis, semiconductors, nanotechnology, sensors, ionic liquids, surface chemistry, electrochemistry, instrumental science, photochemistry, and more.

Dr. Abdullah Mohamed Asiri received a Ph.D. from the University of Wales, College of Cardiff, UK, in 1995. He has served as the head of the Chemistry Department at King Abdulaziz University, Jeddah, Saudi Arabia, since 2009. He is also the founder and director of the Center of Excellence for Advanced Materials Research (CEAMR) and a professor of Organic Photochemistry. His research interests include color chemistry, synthesis of novel photochromic and thermochromic systems, synthesis of novel coloring matters and dyeing of textiles, materials chemistry, nanochemistry, nanotechnology, polymers, and plastics. He is the editor-in-chief of King Abdulaziz University Journal of Science. He is also a member of the editorial board of Pigments and Resin Technology, Organic Chemistry in Sight, and Recent Patents on Materials Science. Dr. Asiri is the vice-president of the Saudi Chemical Society (Western Province Branch).

Dr. Anish Khan is an assistant professor in the Chemistry Department, Center of Excellence for Advanced Materials Research (CEAMR), Faculty of Science, King Abdulaziz University, Jeddah, Saudi Arabia. He obtained a Ph.D. from Aligarh Muslim University, India, in 2010. He has research experience in the field of synthetic polymers and organic-inorganic electrically conducting nanocomposites. Dr. Khan completed postdoctoral studies at the School of Chemical Sciences, University Sains Malaysia (USM), in electroanalytical chemistry (2010–2011). He has years of research and teaching experience and has published more than 127 research papers in refereed international journals, more than twenty international conferences/workshops, five books, and fourteen book chapters. His research interests include preparation and characterization of hybrid nanocomposite materials and their applications, polymeric inorganic cation-exchange materials, electrically conducting polymeric materials, composite materials as sensors, green chemistry, heavy metals, and biosensors.

Dr. Inamuddin is currently an assistant professor in the Chemistry Department, Faculty of Science, King Abdulaziz University, Jeddah, Saudi Arabia. He has extensive research experience in multidisciplinary fields of analytical chemistry, materials chemistry, and electrochemistry, and, more specifically, renewable energy and the environment. He has published 127 research articles in international journals of repute and eighteen chapters in books published by renowned international publishers. He has also edited thirty-nine books. Dr. Inamuddin is a member of various journal editorial boards and is an associate editor for journals such as Environmental Chemistry Letter, Applied Water Science, Euro-Mediterranean Journal for Environmental Integration, and Springer-Nature. He is also an editor for the Eurasian Journal of Analytical Chemistry.

Thamer Tabbakh is an assistant professor in electrical engineering and optics and photonics. He is currently the head of the Labs and Equipment Administration for King Abdulaziz City for Science and Technology (KACST), Riyadh, Saudi Arabia. Dr. Tabbakh obtained advanced degrees from the University of Central Florida College of Optics and Photonics (CREOL-UCF), which is one of the top ten optics and photonics colleges in the United States.

Contents

Preface

Post-Transition Metals contains recent research on the preparation, characterization, and potential applications of post-transition metals such as gallium, indium, tin, thallium, lead, and bismuth, among others. Interest in the chemistry of post-transition elements has increased significantly in the last two decades. In particular, research on the metals of Group 13 has led to the synthesis and application of some very novel molecules, with implications for organometallic synthesis and new materials development for chemical, biological, medical, and environmental uses. This book also discusses new facts, developments, and applications in the context of more general patterns of physical, structural, morphological, and optical behaviors. Particular attention is paid to the main growth areas, including the chemistry of lower formal oxidation states, cluster chemistry, device fabrication, the investigation of solid oxides and hydroxides, advances in the formation of hybrids with II–V and related compounds, the chemical significance of Group 13 metal complexes, and the growing importance of the metals and their compounds in the mediation of inorganic reactions.

In Chapter 1, Prof. Jana et al. discuss indium oxide-based nanomaterials and their fabrication strategies, properties, applications, challenges, and future prospects. The authors highlight synthesis strategies for indium oxide-based bulk nanomaterials with variable morphologies starting from spherical nanoparticles to nanorods, nanowires, nanoneedles, nanopencils, nanopushpins, and more. In addition, the chapter examines thin-film deposition and periodic 1-dimensional/2-dimensional surface texturing techniques for indium oxide-based nanostructured thin films with regard to their functional properties and applications. The chapter also includes a state-of-the-art survey on fabrication strategies and recent advancements in the properties of indium oxide-based nanomaterials with their different areas of applications.

In Chapter 2, Prof. Revaprasadu et al. report on indium chalcogenide nanomaterials, which are at the forefront of recent technological advancements. There has been an increasing trend in the exploitation of indium chalcogenides in various applications ranging from water-splitting reactions in renewable energy to degradation of dyes in environmental rehabilitation. This trend is attributed to the interesting and unique properties of indium chalcogenide nanomaterials, which can be easily tuned via engineering of particle size, shape, and morphology. In this chapter, the authors outline the preferred attributes of indium chalcogenide nanomaterials that are deemed suitable for certain applications. Furthermore, they explore recent reaction protocols that have been reported to yield good quality indium chalcogenide nanomaterials of multinary configurations (e.g., binary and ternary compounds). Finally, the authors address the urgent need for alternative synthesis routes, such as the use of low-temperature decomposing single-source molecular precursors, to be improved and incorporated in the fabrication of functional nanodevices.

In Chapter 3, Dr. Svirsky et al. discuss the ionic state of indium in perchlorate solutions as well as the physicochemical properties of indium perchlorate. In perchlorate solutions, indium (III) cations attach a larger number of layers of water as well as varying degrees, affect its state and through it, the action spreads further to the property and the boundary medium Thus, this causes either a de-structuring effect or a change in the state of chemical bonds of chlorate ions in concentrated solutions. Thus, the chlorate ion, being structure-forming and exhibiting a proto-acceptor ability, in solutions of group III perchlorates forms exclusively solvate-separated ion pairs due to the high enthalpy of hydration of the corresponding metal cations.

In Chapter 4, Associate Prof. Benabderrahmane investigates the effect of indium oxide on the properties of indium particles, which are used as silicon nanowire catalysts. The author examines the elaboration of indium particles by different annealing processes such as rapid thermal annealing as well as conventional processes. The elaborated particles are dedicated for use as catalysts for growing silicon nanowires growth via the vapor–liquid–solid process. After conventional annealing, the indium layer is broken up into elongated and inhomogeneous islands of micrometric sizes. The annealing conditions influence the catalyst morphology and, consequently, grow silicon nanowires.

In Chapter 5, Prof. Tabbakh et al. report on the elastic, optical, structural, and transport properties of gallium arsenide (GaAs). These excellent properties have led to the production of new and unique devices like high-efficiency light emitters, light sensors, and high-speed switching devices. GaAs is considered an outstanding member of the III–V semiconductor family. It has many exceptional features, especially for use in opto-electronic and micro-electronic devices.

In Chapter 6, Mr. Rouf et al. carry out a comparative analysis of the nanoindentation technique. Nano-indentation is a dynamic perceptible method for attaining mechanical properties from limited content. In delicately regulated tests in which the acceptance of the elastic contact analysis is met, the accuracy of a few percentage points is smoothly obtainable for indentations as micro as 10 nm. Specialists must be constantly aware of the holdings of variations from these suppositions on nano-indentation results. An exact evaluation for load, displacement, and machine concurrence is required, as is an effective rational sketch of the shape of the tip and a configuration devised to reduce the consequences of thermal drift and plasticity. Nanoindentation uses an indirect method of determining the contact area, as the depth of penetration is measured in nanometers, while in conventional indentation the area in contact is measured by elementary measurement of the residual area after the indenter is removed from the specimen. Dynamic hardness is the best result of dynamic indentation, which can be expressed as the ratio of energy consumed during a rapid indentation to the volume of indentation. The parameters taken into consideration are indentation depth, contact force, contact area, and mean contact pressure.

In Chapter 7, Prof. Yamaguchi describes a solar cell developed with GaAs material. The author reviews the progress in III–V compound single-junction solar cells such as Gallium Arsenide (GaAs), Indium Phosphide (InP), Aluminium gallium arsenide (AlGaAs), and Indium Gallium phosphide (InGaP) cells. Results show that GaAs solar cells have 29.1% under the sun, which is the highest ever reported for single-junction solar cells. In addition, the author presents analytical results for non-radiative recombination and resistance losses in III–V compound solar cells by considering fundamentals for major losses in III–V compound materials and solar cells. Because the limiting efficiency of single-junction solar cells is 30%–32%, multi-junction solar cells have been developed. GaP/GaAs–based 3-junction solar cells are widely used in this space. Additionally, the III–V compound solar cells have contributed as space and concentrator solar cells and are expected to be used in large-scale electric power systems and solar cell-powered electric vehicles.

In Chapter 8, Ph.D. Baron et al. report on opto- and nanoelectronic applications with GaAs compounds heteroepitaxy on silicon. The authors show how to overcome the different challenges associated with heteroepitaxy and integration of III-As onto a silicon platform. They present solutions to get rid of antiphase domains for GaAs grown on exact Si(100). To reduce the threading dislocations density, efficient ways based on either insertion of InGaAs/GaAs multilayers defect filter layers or selective epitaxy in cavities are implemented. All these solutions allow for the fabrication of electrically pumped laser structures based on InAs quantum dots active region, which is required for photonic and sensing applications.

Finally, in the last chapter, Dr. Pradhan et al. investigate IMPact ionization Avalanche Transit-Time (IMPATT) diodes based on GaAs for millimeter-wave applications with reference to silicon. The chapter presents DDR IMPATTs based on GaAs designed to operate at mm-wave window frequencies of 94, 140, and 220 GHz. Both the DC and small-signal performances of these devices are investigated by using a small signal simulation technique developed by the authors. The efficiency, output power, and power density of a GaAs IMPATT are greater than that of a Si IMPATT. Results show that the DDR IMPATTs based on GaAs are most suitable for generating radio frequency power with maximum conversion efficiency up to 220 GHz. This chapter looks at the benefits of GaAs in power electronics applications, reviews the current state of the art, and shows GaAs is a strong and feasible candidate for IMPATTs. It is also well known that at a given frequency the microwave and mm-wave power output of an IMPATT diode are proportional to the square of the product of the semiconductor critical field and carrier saturation velocity. For mm-wave frequencies greater than 94 GHz, a GaAs semiconductor is the best choice for fabricating a DDR IMPATT device.

Dr. Mohammed Muzibur Rahman
Center of Excellence for Advanced Materials Research
and Department of Chemistry,
Faculty of Science,
King Abdulaziz University,
Jeddah, Saudi Arabia

Dr. Abdullah Mohamed Asiri, Dr. Anish Khan and Dr. Inamuddin
King Abdulaziz University,
Kingdom of Saudi Arabia

Thamer Tabbakh
Assistant Professor,
King Abdulaziz City for Science and Technology,
Saudi Arabia

Chapter 1

Indium Oxide Based Nanomaterials: Fabrication Strategies, Properties, Applications, Challenges and Future Prospect

Hasmat Khan, Saswati Sarkar, Moumita Pal, Susanta Bera and Sunirmal Jana

Abstract

Nanostructured metal oxide semiconductors (MOS) in the form of thin film or bulk attract significant interest of materials researchers in both basic and applied sciences. Among these important MOSs, indium oxide (IO) is a valuable one due to its novel properties and wide range of applications in diversified fields. IO based nanostructured thin films possess excellent visible transparency, metal-like electrical conductivity and infrared reflectance properties. This chapter mainly highlights the synthesis strategies of IO based bulk nanomaterials with variable morphologies starting from spherical nanoparticles to nano-rods, nano-wires, nano-needles, nanopencils, nanopushpins etc. In addition, thin film deposition and periodic 1-dimensional (1D)/2-dimensional (2D) surface texturing techniques of IO based nanostructured thin films *vis-à-vis* their functional properties and applications have been discussed. The chapter covers a state-of-the-art survey on the fabrication strategies and recent advancement in the properties of IO based nanomaterials with their different areas of applications. Finally, the challenges and future prospect of IO based nanomaterials have been discussed briefly.

Keywords: metal oxide semiconductor, indium oxide based nanomaterials, fabrication strategies, periodic surface texturing, advanced applications

1. Introduction

It is no doubt that nanomaterials have attracted significant attention for both basic and applied sciences because these materials in nanodimension (1–100 nm) exhibit novel features including high surface area, excellent physical and chemical stability and lower material density compared to their bulk counterpart. In fact, these features of the nanomaterials help the researchers to design and fabricate novel functional nanomaterials/devices for practical use. Today, various forms of nanomaterials such as quantum dots, nanoparticles, nanoflakes, nanobelts, nanoribbons, nanosheets, nanofilms, nanotubes, nanofibers even nanocomposites have widely been used to improve the materials properties including thermal, electrical, mechanical, optoelectronics, corrosion resistant, self-cleaning, and sensing [1–3].

Over the past decades, nanostructured metal oxide semiconductors (MOSs) have drawn tremendous attention to materials researchers due to their widespread applications in various fields [4–6]. Among various MOSs, indium oxide (In_2O_3) has been investigated widely owing to its wide band gap, high electrical conductivity, stability and excellent optoelectronic properties [2, 7]. In_2O_3 (IO) is a wide band gap n-type semiconductor with direct band gap energy of 3.6 eV at room temperature [7, 8]. It is found that the band gap energy of IO thin film primarily depends on various factors such as annealing temperature and atmosphere as well as the nature of the substrate on which the film is to be deposited. The growth temperature also influences upon the morphological, structural, electrical and optical properties of IO based thin films. It is to be further noted that the thin films with enhanced functional properties like high electrical conductivity and visible transparency can be achieved by controlling the annealing temperature and atmosphere during the fabrication process [9]. It is noteworthy that these materials are suitable for different applications such as photovoltaic devices, liquid crystal displays, transparent conductive electrode in electronic devices, solar cells and flat panel displays, photodetectors, gas sensors, heat reflecting windows etc. [2, 8, 10]. On the other hand, different nanostructured IO based bulk nanomaterials such as nanosheets, nanowires, nanoparticles, quantum dots, single crystals are found to have potential applications [8, 11–13].

In the last decade, IO/IO based nanomaterials has been studied extensively. Around 54 years ago, Groth *et al.* [14] demonstrated that the small amount of Sn or Ti doping into IO can significantly enhance the electrical conductivity and infrared reflectivity without losing optical transparency in visible region. Based on this experimental observation, the Sn-doped In_2O_3, popularly known as indium tin oxide (ITO) creates an active area of research and development in the field of electrochromic and infrared reflective windows, light emitting diodes, transparent contacts for solar cells and flat panel displays and cladding layers for InGaN-based lasers [2]. Recently, the rapid increase in production of various electronic/optoelectronic devices with ITO results a sharp increase in price of indium. In order to minimize the cost without sacrificing the functional properties, indium oxide based thin films have been fabricated [15–17]. In this respect, the formation of heterostructure with band gap engineering of IO or IO based nanomaterials improve its functional properties for advanced applications especially in transparent electronic devices and sensors [18–20]. In this regard, Wang *et al.* [18] reported hierarchically structured ZnO decorated with IO nanoparticles synthesized by one-pot sol-gel process towards improvement in n-butanol sensing performance. Moreover, N-doped graphene quantum dots modified three-dimensional ordered macroporous IO based nanocomposites had been fabricated for NO_2 gas sensing application [10]. On the other hand, IO based nanomaterials are largely used for microelectronics and optoelectronic applications [2, 7, 19] and also found in significantly improved stability of solar cells without negotiating the performance of IO/ZnO electron transporting bilayer, synthesized by solution-process as reported by Kirmani *et al.* [20].

There are various methods now available to synthesize different types of IO/IO-based nanomaterials. The common techniques to deposit the thin films of IO based nanomaterials are sol-gel, spray pyrolysis, Ink-Jet printing, physical/chemical vapor deposition and atomic layer deposition [2, 4, 8, 21]. On the other hand, IO-based nanomaterials are generally synthesized by sol-gel, solvothermal/hydrothermal, co-precipitation, thermal evaporation and solid state reaction methods [22].

This chapter mainly highlights the synthesis strategies of IO based bulk nanomaterials with variable morphologies starting from spherical nanoparticles to nano-rods, nano-wires, nano-needles, nanopencils, nanopushpins etc. In addition, thin film deposition and periodic 1D/2D surface texturing techniques of IO based nanostructured thin films *vis-à-vis* their functional properties and applications have been discussed. Thus, the chapter covers a state-of-the-art survey on the fabrication strategies and recent advancement in properties of IO based nanomaterials with their different areas of applications. Finally, the challenges and future prospect of IO based nanomaterials have been briefly discussed.

2. Nanostructured metal oxide semiconductors

Now-a-days, nanostructured MOSs draw special attention owing to their promising applications in various areas such as electronic, optoelectronic, energy storage and conversion, adsorption, catalysis and sensing for their fascinating characteristics including high surface-to-volume ratio, surface permeability, light harvesting capability, electrochemical and photochemical properties [4–6, 23]. Nanostructured materials can be classified as zero-dimensional (0-D, nanoparticles, core-shell nanoparticles), one-dimensional (1-D e.g. rods/wires), two-dimensional (2-D e.g. layered structures, composite nanowires), and equiaxed or three dimensional (3-D e.g. nanotubes/nanowires bundles). It is to be noted that hierarchical nanostructures can be formed by combining 0-D, 1-D, 2-D and 3-D nanostructures [24]. These nanoscale structures of MOSs are capable to exhibit an improvement in mechanical, optical, electronic, optoelectronic or magnetic properties [1, 24].

2.1 Indium oxide based nanomaterials

In the next sub-sections, a discussion has been made on IO based nanomaterials especially bulk nanomaterials and porous nanomaterials including nanostructured thin films.

2.1.1 Bulk nanomaterials

Development of functional nanomaterials in bulk form can fulfill the purpose of achieving some special properties which can not be possible in the form of thin film/coating. This is because the properties such as structural, optical, optoelectronic, microstructural, electrical etc. of a material in bulk form can greatly differ from its thin film counterpart. Thus, the fabrication of bulk nanomaterials is also highly essential for their widespread applications. In this respect, the method of their syntheses can determine the specific structural features related to grain size, interface boundaries, porosity, structural defects and so on [25–27]. In this regard, for the synthesis of some indium oxide based bulk nanomaterials few well-established methods and the applications of the products are listed in **Table 1**.

Nanomaterials with porous architecture are very much important in the field of nanoscience and nanotechnology because of the ability of the materials to interact with atoms, ions and molecules not only at their surface but also throughout the bulk region. Moreover, the surface area which is mainly dependent on the particle size, shape and volume of the void space present in a porous nanomaterial is directly related to the functional property [4–6]. Thus, to obtain superior functional properties, the textural properties should be tuned accordingly.

System	Synthesis method	Ref.
NiS-IO-GO	Ultrasonic/hydrothermal	[28]
WO_3-IO	Sol-gel	[29]
SnO_2-IO	Precipitation	[30]
Y_2O_3-IO	Co-precipitation/sol-gel	[27]
Mn(II) doped IO foam	Sol-gel	[31]
Organic–inorganic IO foam	Sol-gel	[32]
Colloidal IO nanoparticles	Laser ablation	[33]

Table 1.
Synthesis and application of IO based bulk nanocomposite materials.

2.1.2 Nanostructured thin films

Indium oxide based nanostructured thin films have great significance owing to their variable band gap energy (3.2–3.8 eV) with high visible transparency, substantial environmental and chemical stability as well as high electron mobility and metal-like electrical conductivity [34–45]. These nanostructured thin films have been fabricated (**Table 2**) with excellent optical and electrical properties towards various applications [34–45]. However, the thin films are mostly used as TCOs [34–45]. In this regard, ITO thin film is known to be one of the most extensively used TCO. In addition, after modification of thin film surface (**Figure 1**) by periodic texturing adopting soft lithography or breath figure process (BRF), the surface textured films can be used for light frequency modulation, photoelectrochemical application and photocatalysis [4].

Thin film system	Deposition method	Application	Ref.
Indium oxide	Spin coating	Buffer layer and transparent electrodes in solar cells	[34]
Indium-doped ZnO	Spin coating	High-performance thin film transistors (TFTs)	[35]
Indium-gallium-zinc oxide	Ink-jet printing	High performance printed TFTs	[36]
InGaZnO	Spin coating	White light illumination and TFT channel	[37]
Zinc indium oxide	Sol-gel and breath figure	Photoelectrochemical water splitting	[4]
Antimony doped indium oxide	Sol–gel	Heat absorbing window glass fenestration	[38]
$Cu(In,Ga)(S,Se)_2$ (CIGSSe)	Spray pyrolysis	Solar cell	[39]
Indium zinc oxide	Inkjet-printing	TFTs	[40]
ITO	PVD with glancing angle	TCO, gas-sensors, self-cleaning	[41]
ITO	Ultra-thin RF magnetron sputtering	Top electrode in photovoltaic devices	[42]
Indium-gallium-oxide	CVD process	Ultraviolet phototransistors	[43]
Indium oxide	Atomic-layer deposition	TFTs	[44]
ITO	Sputtering	Acetaldehyde sensing	[45]

Table 2.
Different methods of deposition and applications of selected IO based nanostructured thin films.

Figure 1.
FESEM images of IO based nanostructured thin films: (a) zinc indium oxide, (b-d) ITO, (e) indium gallium zinc oxide and (f) 1D surface patterned zinc indium oxide [4, 46–49]. (Copyright reserved to the American Chemical Society (2017), AIP Publishing (2015), Springer Nature (2018) and AIP Publishing (2014) for references [4, 46, 48, 49], respectively).

3. Fabrication strategies of indium oxide based nanomaterials

3.1 Indium oxide based thin films

Indium oxide based nanomaterials including nanostructured thin films and bulk nanomaterials are found to be a highly exploration area in the nano domain [34–45]. For deposition of nanostructured IO based thin films, several techniques including sol-gel, sputtering, physical vapor deposition (PVD), chemical vapor deposition (CVD), ink-jet printing, spray pyrolysis, atomic layer deposition (ALD) are available [34–45]. These are discussed in the next sub-sections.

3.1.1 Sol-gel coating technique

Sol-gel coating technique such as dip coating, spin coating or sometimes drain coating (especially for large size and heavy weight substrate) is also a very much useful for development/fabrication of various functional nanostructured thin films that are suitable for the use especially in the field of microelectronics and optoelectronics [4, 34–38]. The coating technique can fulfill to obtain desired physical and optical properties of the nanomaterials [34–38]. In this respect, sol-gel coating technique as a facile fabrication strategy has already been established as cost effective one for deposition of nanostructured IO based thin films [34–38]. Among other thin film fabrication techniques such as sputtering, CVD, ALD, PVD, spray pyrolysis and ink-jet printing, sol-gel technique is become a convenient one where high chemical and environmental stabilities of the nanomaterials can be obtained [34–38]. Sol-gel coating techniques can be applicable to deposit a huge numbers of high-performance nanostructured IO based thin films including ITO, IZO, IZGO, ZIO, IAO, Sb-doped IO, Cd-ITO, Cr-ITO for various applications [34–45]. A schematic diagram as shown in **Figure 2** where IO based thin films by sol-gel coating technique is described. In addition, some sol-gel based nanostructured IO thin films with their area of applications are highlighted in **Table 2**.

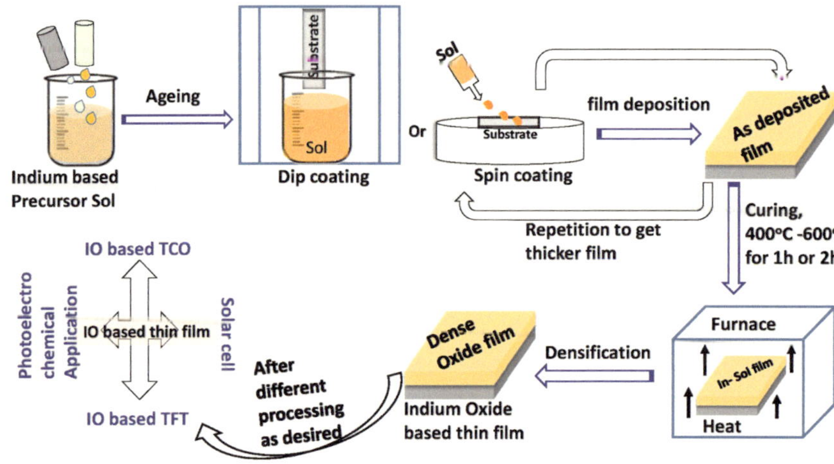

Figure 2.
Schematic presentation for the fabrication of IO based thin films by sol-gel coating technique.

3.1.2 Spray pyrolysis

In spray pyrolysis technique for IO based thin-film deposition includes spraying of suitable metal salt solution onto a heated substrate [39]. In this technique, the main steps are precursor solution atomization, transportation of resultant aerosol and decomposition of precursor onto a substrate [39]. Another important factor of this technique is to select an appropriate type of atomizer for desired application. Important parameters that have to be controlled are atomization, solute concentration, temperature gradient and carrier gas [39]. Adopting this technique, the fabrication of different IO based nanostructured thin films have been reported [2, 39]. Some of the potential nanostructured thin films and their applications are given in **Table 2**.

3.1.3 Ink-Jet printing

Ink-Jet printing technique (IJP) can mainly be divided into nozzle based digital inkjet printing and non-digital screen, offset, flexography and gravure printing [40]. In this technique with appropriate ink solvents, a widespread range of flexible thin-film devices can be fabricated such as transistors, light-emitting devices, sensors and energy harvesting and storage devices [40]. Although, gravure printing results high-throughput with high resolution and noble pattern fidelity but due to the characteristic contact nature of gravure printing along with the use of high-viscosity ink with binders, the contamination/residue related issues and degradation of printing materials may occur [40]. Few reports with respect to IO based nanostructured thin films are listed in **Table 2**.

3.1.4 Sputtering physical vapor deposition

A commonly used method for IO based thin film deposition is PVD in which the coating generates onto a substrate through atom by atom [41]. The PVD involves the atomization or vaporization of material from a solid source called target [41]. In this technique, the substrate majorly influences the properties of thin film. It is worthy to note that the deposition method must be performed under vacuum, plasma, gaseous or electrolytic environment. In this technique, the stresses generated into a thin film during cooling process or melting of substrate (mostly for an organic polymer) can limit the deposition process [41, 45]. Transparent conducting ITO thin films over silicon wafer can be fabricated by PVD techniques such as magnetron sputtering, vacuum evaporation, ion-plating [41, 45] towards gas-sensors, SERS, electrochromism and self-cleaning applications [41, 45].

3.1.5 Chemical vapor deposition

The IO based nanostructured thin films can also be fabricated by CVD [43]. In this process, the combination of gases react with substrate surface at comparatively high temperature that leads to decay of a particular constituent of gas combination. Hence, the fabrication of a metal or composite solid film can be deposited onto a substrate. This process can either be a pyrolysis of vapors of single organometallic compounds or a second reactant as an intermediate in vapor phase [43]. In this regard, some reports are available (**Table 2**) on single-phase metastable rhombohedral ITO epitaxial thin films with high transparency and electrical conductivity. The films had been deposited on Al_2O_3 substrate by CVD [43]. Also, indium gallium oxide thin film had been developed by co-sputtering using Ga_2O_3 and In_2O_3 as targets at room temperature [43].

3.1.6 Atomic layer deposition

ALD is also a technique for deposition of nano structured IO based thin films [44]. This technique mainly based on sequential pulsing principle of precursor chemicals in vapor state where each pulse is almost one atomic layer thin. The excess reactants as by-products can purge or evacuate with an inert carrier gas (e.g. N_2/Ar) [44]. The precursor used in this technique is pulsed into a chamber under vacuum (<1 Torr) condition for a certain period of time during each half-reaction. The process is cycled afterwards until a suitable film thickness reached. By applying this technique, a layer of very high aspect ratio of ITO crystals with nanoporous architecture can be fabricated. These materials can be used in photovoltaic or spectroelectrochemical applications [44]. On the other hand, In_2O_3 TFTs with ALD Al_2O_3 gate dielectrics had already been developed with significantly good electrical performance (e.g. field effect mobility, 7.8 $cm^2\ V^{-1}\ s^{-1}$ and on/off current ratio, 10^7) [44].

3.2 Periodic surface texturing of thin films

Periodic texturing also called patterning on thin film surfaces is a potential technique for the fabrication of photonic nanostructures for various optical applications [50]. It is worthy to note that the improved solar light absorption with high surface to volume ratio and enhanced light harvesting efficiency of the MOS thin films can be enhanced by periodic nanostructuring [6, 50]. There are several surface texturing techniques like conventional photolithography, nano-imprint lithography, electron-beam lithography, laser patterning, dip-pen lithography, reactive ion etching etc. available in the literatures [6] but these techniques are very costly, complicated, time consuming and also have several limitations owing to the nature of the component materials [6, 12]. Hence, versatile, simple and cost effective unconventional soft lithography is used now-days as an alternative to these conventional lithography techniques. This technique is largely used to generate periodic structures on metal oxide/mixed metal oxide including polymer based thin film surfaces [6, 12, 13]. In the next sub-sections, a special emphasis is given on sol-gel based soft lithography technique to perform periodic surface texturing on mixed metal oxide thin films.

3.2.1 Importance of periodic surface texturing

It is no-doubt that periodic surface texturing (pattering) is used to improve the functional properties of metal oxide thin films. The main objective of surface texturing is to effectively manage the incident light into thin film matrix. Thin films with different periodic surface structures are capable to enhance light absorption *via* light scattering and anti-reflective effects [50]. Improved light absorption ability of nanostructure thin films can enhance the performance of optoelectronic devices [6, 50]. Generally, light management in the nanostructure device based on two simple strategies- (a) anti-reflection and (b) enhancement in light absorption [50]. The enhanced light absorption occurs in surface textured thin films through multi-internal reflection which increases the light propagation length into the absorbing layer. Theoretically, it is possible to improve the light absorption up to an enhancement factor of $4n^2$ (Lambertian limit) where 'n' is denoted as refractive index of the material [50]. It is found that ordered three-dimensional nanostructured materials reach or exceed the Lambertian limit. Thus, it is established that the light absorption not only depends on the materials properties but also on the geometry of materials [50].

3.2.2 Periodic surface texturing techniques

As already stated in the previous sub-sections, periodic surface texturing of thin films is generally performed by conventional photolithography. Beside photolithography technique, several other techniques like nano-imprint lithography, electron-beam lithography, laser patterning, dip-pen lithography, reactive ion etching etc. are also used for the surface texturing [6, 13]. In soft lithography, one of the nonconventional lithography techniques, a soft organic material is mostly used to produce patterned structures without using light or any other high energy particles [6]. The main feature of this technique is to use a surface patterned elastomeric stamp which is generally made of polydimethyl siloxane (PDMS). This PDMS stamp can be used either as a mold to impart the patterns through physical confinement of a liquid precursor that dries to build the patterned film or as a stamp to directly transfers the precursor material to the substrate [51]. This technique mainly consists of different types such as replica molding (REM), microcontact printing (μCP), micromolding in capillaries (MIMIC), microtransfer molding (μTM). By using soft lithography techniques, it is possible to fabricate periodic surface textured films with features, ≥30 nm [51].

3.2.3 Applications

Periodic surface textured metal oxide based thin films have diverse applications in various fields like self-cleaning, photovoltaics, catalysis, energy conversion and storage, electronic devices, sensor and solar water splitting [52]. Now-a-days, surface patterned metal oxide thin films are also largely used in photovoltaic cells as active layers, photocatalysis and photoanode in photoelectrochemical (PEC) cells [4–6, 52]. Nanostructuring on the thin film surface increases the active surface area as well as photon capturing ability which are beneficial for the enhancement of photocatalytic and PEC performances [6, 13]. Thus, the PEC performance of MOS thin films can be improved by periodic surface texturing. It is worthy to note that overall PEC performance for solar water splitting depends on three fundamental factors- (i) absorption efficiency (η_{abs}), (ii) charge separation efficiency (η_{sep}) and (iii) charge transfer efficiency (η_{trans}). The performance for solar water splitting is expressed as $\eta_{abs} \times \eta_{sep} \times \eta_{trans}$ [53]. It is very challenging to get high value of the product of η_{abs} and η_{sep} (i.e. $\eta_{abs} \times \eta_{sep}$) because these are coupled with each other [53]. By increasing the active layer thickness, it is possible to increase η_{abs} value but it reduces the η_{sep} value.

As a result, the product of η_{abs} and η_{sep} decreases. However, the nanostructuring on MOS metal oxide thin film surfaces can mitigate the problem. Several nanostructured mixed metal oxide thin films including IO based nanostructured thin films are reported for improving PEC performances [4, 6, 53].

3.3 Synthesis of indium oxide based bulk nanomaterials

Indium oxide/indium oxide based nanomaterials can be synthesized by various synthesis methods. Some of them are sol-gel, solvothermal/hydrothermal, co-precipitation, thermal evaporation and solid state reaction. These are discussed in the next sub-sections.

3.3.1 Sol-gel method

Sol-gel is a wet chemical method for synthesis of various nanomaterials from sols. Generally, this method involves controlled hydrolysis of metal alkoxides like zirconium propoxide or metal salts such as metal nitrates or chlorides in an aqueous or organic solvent medium. It is to be noted that at the initial step of this method the hydrolysis and polycondensation reactions occur that lead to the generation of polymeric/colloidal sol with particles of nano dimension [5, 18]. By increasing mass of the desired material into sol or other significant changes in sol like solvent substitution, pH variation, solvent evaporation, etc. results the formation of gel with three-dimensional network of porous nature. The solvent is enclosed inside the porous gel structure [5, 18]. Heat treatment of the gel structure produces dense ceramics. It is well-known that the extremely pure and homogeneous multicomponent oxide can be synthesized by this method. Metal ions doped indium oxide and indium oxide based nanomaterials in the form of thin films or bulk nanocomposites can be fabricated/synthesized by adopting the sol-gel method [4–6, 12, 13, 18].

3.3.2 Solvothermal/hydrothermal method

Solvothermal or hydrothermal synthesis method is termed depending upon the solvent used in the synthesis method. In the hydrothermal method, water is generally taken as solvent whereas organic solvent instead of water is used in solvothermal method. In these methods, aqueous solutions of metal nitrates, chlorides and acetates are generally used as precursor materials for the synthesis of metal oxides [8, 10]. In this method, the precursor materials and solvents are taken in a particular stoichiometric ratio and stirred for a particular time period to obtain a homogeneous solution. Then, the solution is transferred into a Teflon coated stainless steel autoclave and placed it in an oven at elevated temperature for a certain time. Finally, the autoclave is allowed to cool down at room temperature and the precipitates obtained is dried and cured at higher temperature. Different indium oxide based bulk nanomaterials have been synthesized by solvothermal/hydrothermal method. In this regard, Suzuki *et al.* [54] synthesized indium tin oxide nanoparticles *via* solvothermal method for sustainable coating application.

3.3.3 Co-precipitation and thermal evaporation

Co-precipitation is a simple classical method to synthesize metal oxide nanomaterials. This method is cost-effective, very fast process and useful for larger scale industrial applications [27, 30]. By this method, it is possible to synthesize highly pure nanomaterial through an eco-friendly route. In this typical method, metal salts in the form of nitrate, chloride, or oxychloride as precursor materials are generally dissolved in aqueous solution and precipitated these into their corresponding hydroxides by addition of a base like sodium hydroxide or ammonium hydroxide. Finally, the precipitates are washed and calcined at high temperature to get metal oxide nanomaterials. Several reports are available on synthesis of indium oxide based nanomaterials by this method [55, 56].

3.3.4 Solid state reaction

Solid-state reaction is a well-known method for the synthesis of polycrystalline material from solid precursor materials. Generally, the reaction occurs at very high temperature. The main advantage of this method is its simplicity and the ability of large scale industrial production. Synthesis of pure In_2O_3 nanoparticles had been performed via solid state reaction method for the fabrication of optoelectronic devices by Jothibas *et al.* [57]. Moreover, Bykova *et al.* [58] reported Co-In_2O_3 nanocomposites thin film by solid-state reaction method and investigated its structural and magnetic properties. Recently, Co-In_2O_3 nanocomposites and cobalt-doped In_2O_3 have attracted significant attention due to their applications in optoelectronic, spintronic devices and gas sensors [58].

4. Properties of indium oxide based nanomaterials

4.1 Indium oxide based nanostructured thin films

IO based nanostructured thin films are immensely important due to their excellent optical, electrical, and mechanical properties suitable for various applications like energy conversion, biological and chemical sensing, solar cells, thin film transistors etc. [34–45]. Structural, optical, magnetic and electrical properties of IO based nanostructured thin films are discussed in the next sub-sections.

It is well known that IO based nanostructured thin films can act as excellent n-type transparent conducting oxides (TCOs) [34–45]. In bcc-Sn doped In_2O_3 forming ITO, the low formation energy implies a greater abundance of both the neutral and the cationic states of Sn dopant [38]. The structures of the thin films as confirmed by X-ray diffraction study, indicate that the films are polycrystalline with bcc structure having a 100 intensity peak at (222) plane of the crystal lattice. It is observed that the mobility of atoms and clusters on the surface of a substrate is proportional to their energy that would increase with increasing curing temperature. This would lead to the growth of In_2O_3 crystallites along a crystal plane (100). Among the fabricated IO based thin films, ITO is the most widely used efficient TCO due to its low energy of defect formation towards enhancing greater electrical properties [34–45]. It is worthy to note that the optical properties of IO based thin films are primarily dependent upon post annealing temperature, film microstructure, film physical thickness, surface roughness, levels of impurities, defect (like oxygen vacancies) concentration and deposition parameters [16, 17, 34–45, 59]. Low absorption (0.04–1.10%) of incident light in visible region is a crucial factor for IO based TCOs. The optical band gap widening or narrowing of IO based films occurs also based on dopant concentration [34]. In this context, different values as obtained from the reported works on IO based thin films show a high optical transparency (82–93%) of the films [34–45]. On the other hand, it is very much important to achieve the magnetic properties of IO or IO based thin films and bulk nanomaterials. In order to obtain the magnetic properties of these nanomaterials different magnetic metal such as Cr, Mn, Fe, Co, etc. ions with variable oxidation states are generally doped into the metal oxide [59]. It is found that the magnetic property is induced in indium oxide due to the presence of oxygen vacancies. The property basically depends on the nature/concentration of external dopant [59, 60].

4.2 Indium oxide based bulk nanomaterials

Functional properties of bulk nanomaterials can sometimes be advantageous compare to nanostructured thin films and the properties such as structural, optical, optoelectronic, microstructural and electrical in bulk nanomaterials differ from their respective thin film counterpart. In this respect, the preparation methods that govern the generation of specific structural features, porosity and defects in the bulk nanomaterials. It is reported that ITO and nickel doped ITO nanomaterials exhibit X-ray diffraction peaks similar to that of pure In_2O_3 with cubic bixbyite structure [27–30]. To calculate the band gap energy (BGE), the optical reflectance spectrum of a bulk nanomaterial can be recorded and then converted into its absorption spectrum. The absorption coefficient (α) can be determined using Kubelka-Munk function relation, $\alpha = (1 - R)^2/2R$. The BGE (E_g) can be determined from the relation, $\alpha h\upsilon = A\,(E_g - h)^{1/2}$, where h, υ and A are Plank's constant, frequency of light, proportionality constant, respectively [57, 61]. The BGE value of a metal doped IO/IO based nanomaterial varies depending on the nature of dopant element [27–30]. On the other hand, magnetic properties of IO/IO based nanomaterials are of great interest for basic science. Pure In_2O_3 shows diamagnetic behavior while the ITO displays ferromagnetism at room temperature. Moreover, different transitional metal doped In_2O_3 nanomaterials exhibit ferromagnetism at room temperature due to presence of oxygen vacancies [59, 60].

5. Applications of indium oxide based nanomaterials

Because of high optical transmittance and excellent electrical conductivity, indium oxide/IO based nanomaterials have variety of applications especially for fabrication of optoelectronic and microelectronics devices [2, 16, 17]. These are discussed in the next sub-sections.

5.1 Transparent conducting oxide

Generally, transparent conducting oxides (TCOs) are the materials that possess two major properties of (a) high electrical conductivity and (b) excellent optical transparency. Indium oxide is an important MOS material that mainly uses as TCO [62]. The properties of In_2O_3 are improved by doping or coupling of other semiconductors with wide BEGs. Improved electrical and optical properties of In_2O_3 based nanomaterials can promote to fabricate a variety of potential modern devices such as touch screen displays, low emissivity windows, solar cells, and gas sensors [2, 62]. Till date, ITO is the most successful TCO in terms of optoelectronic properties used commercially. It is found that ITO demonstrates a very high electrical conductivity and carrier concentrations without losing visible light transparency [62, 63].

5.1.1 Transparent conductive electrode

Transparent conductive electrodes made with TCOs are hugely used in flat panel displays, touch panels, lamps and thin film solar cells [62, 63]. Among the various TCOs, IO and ITO are highly used to fabricate transparent conductive coatings. It is noted that various properties of TCO thin films can be tuned by selecting suitable dopant with its optimized concentration. Thus, impurity doped ZnO, In_2O_3, and SnO_2 thin films are the best materials for practical utilization as transparent conducting electrodes [62, 63]. Also, different indium oxide based materials have already been explored to fabricate transparent conductive electrode [62–64].

5.1.2 Antireflection coating

Antireflection coating (ARC) is an optical coating which reduces undesirable reflections from substrate surfaces and increases transmittance [65, 66]. It is extensively used in industrial applications such as solar cells and photovoltaics, electronic device displays, and also general purposes, like spectacle and photographic lenses [65–67]. In this regard, IO based nanomaterials especially ITO has gained a special attention and ARC can be applied on this TCO thin films for the applications [66, 67].

5.1.3 Infrared reflective and electrochromic coatings

In electrochromic devices, the optical properties such as absorption, transmission, reflection and/emission can be changed when an electrical potential is applied. These devices are ubiquitous in daily life and are used as antidazzle rearview mirrors in cars as well as state-of-charge indicator strips of batteries. Now-a-days, an aim of research in electrochromic devices is to develop smart windows. It is applied in buildings to save energy cost by controlling the incident sun light and heat radiation. In this regard, Llordes *et al.* [68] reported a nanocomposite capable of tunable visible light as well as near-infrared transmittance. It is found that the development of smart windows with various switchable states including transparent, dimmed, cold and hot is possible with the use of this material. In this nanocomposite, the heat radiation transmittance can be controlled by ITO nanocrystals that exhibit an extensively tunable localized surface plasmon resonance [69]. It is also reported that the infrared transmittance is electrochemically controlled by the carrier concentration of ITO nanocrystals [69].

5.2 Optoelectronic applications

5.2.1 Photovoltaic cells

In photovoltaic (PV) cells/solar cells sunlight is directly converted into electricity. Generally, a PV cell is fabricated by two or more thin layers of semiconducting material. On illumination of sun light, the PV cells generate electrical charges that are conducted away by metal contacts. The PV cells have minimal maintenance cost and have a long life. Another important advantage of PV cells is that they generate solar electricity without emitting greenhouse or any other environmentally hazardous gases. Single PV cells generally provide very small amount of current. In order to obtain a demandable current and voltage output, a number of PV cells are connected together in series and confined with a glass cover, called solar cover glass and plastic sheet to form a PV panel.

Basically, it is a p-n junction diode. Under the exposure of light into the p-n junction, number of electron-hole pairs are generated and separated to produce electricity [70]. Up to date, three types of photovoltaic cells are available such as the first generation, second generation, and third generation PV cells. Crystalline silicon wafers as p-n junction diodes are the first generation cells. It is noted that the silicon solar cells has better efficiency but these are very expensive. The second generation solar cells are based on thin films of crystalline or amorphous silicon and $CuInSe_2$-based cells. It is found that the third generation cells such as polymer-based solar cells, nanocrystals based solar cells, dye-sensitized solar cells, quantum dot sensitized solar cells, perovskite solar cell and concentrated solar cells are very potential to harvest the solar energy (**Figure 3**) [70]. Recently, the inexpensive and flexible polymer thin films with stable inorganic nanostructures as fourth generation solar cells are developed to improve the efficiency [70]. In this regard, different indium oxide based nanomaterials are used for fabrication of advanced solar cells that can efficiently convert light energy into electricity [71, 72].

5.2.2 Photoelectrochemical cell

Photoelectrochemical (PEC) cell is a typical device where solar energy is converted into chemical energy in the form of fuel. Generally, it is made with photoactive semiconductor electrodes (photocathode and photoanode). The electrodes are immersed in a suitable electrolyte solution and the semiconductor-electrolyte junction is illuminated with a light source that has higher energy compared to the BGE of the semiconductor (**Figure 4**). As a result, the electrons and holes are

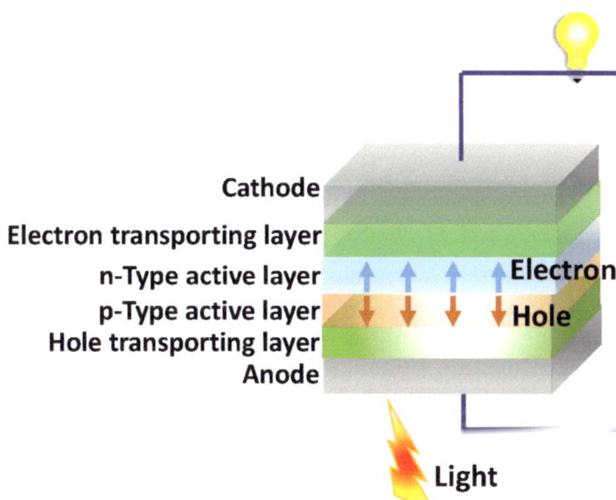

Figure 3.
Schematic presentation of a solar cell.

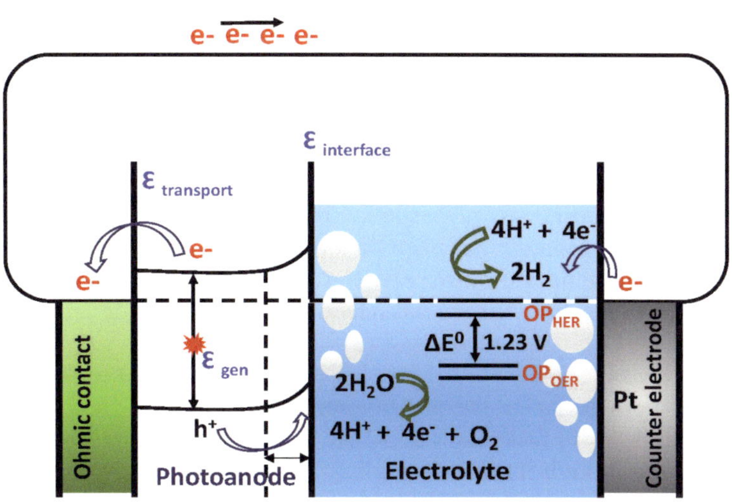

Figure 4.
Schematic diagram displaying the basic principle with key parameters of PEC water splitting.

generated and separated in the space charge region [4, 6, 11]. Now-a-days, various indium oxide based nanomaterials are used as both photoanode and photocathode materials for PEC water splitting application. Under light illumination, the photogenerated minority carriers (holes) in photoanode reach at the interface of electrolyte-semiconductor whereas majority carriers (electrons) accumulate at the interface of semiconductor-conducting substrate and transported with the help of a connecting wire to the counter electrode. These photogenerated charge carriers react with electrolyte solution to produce O_2 and H_2 [73]. In this regard, Cao *et al.* [11] fabricated a 3D hierarchically porous In_2O_3/In_2S_3 heterostructure array onto fluorine-doped tin oxide glass substrate *via* an ion exchange–induced synthesis and used the heterostructure film as photoanode in PEC cell with incident photon-to-current conversion efficiency of 76% at 400 nm.

5.2.3 Photodiodes

Photodiode is a lightweight sensor in which light energy is converted into electrical current or voltage. It is made of semiconducting materials and p-n junction is developed within it. Generally, it accepts light energy as input to produce an electric current as output. Different indium oxide based materials have been utilized as photodiode applications [74, 75].

5.3 Other applications

5.3.1 Photocatalysis

IO based photocatalysts are used for removal of volatile organic compounds, degradation of organic pollutants, hydrogen evolution and so on [76]. The photocatalytic activities of a single photocatalyst is usually limited due to their high recombination rate of photo-generated charge carriers and also their low utilization of visible light energy. In this respect, the fabrication of semiconductor nanocomposite based photocatalysts by imposing various novel strategies (such as doping impurity element into the metal oxide semiconductor or coupling with other semiconductor oxides, metals, and carbon) have been investigated as a feasible and promising strategy to overcome the shortcomings. Some IO based photocatalysts that are already studied by several researchers are displayed in **Table 1**.

5.3.2 Gas sensors

Gas sensors have widely been explored in recent years to monitor and rapidly detect flammable, explosive, and toxic gases in an environment. The most important factors in determining the gas-sensing performance of these sensors are sensitivity, working temperature, response/recovery time, and also the selectivity. Thus, significant research has been focused on exploring various methods to lower the working temperature, increase the sensitivity, shorten the response/recovery time, and also to improve the selectivity of metal oxide semiconductor based sensors. In the last decades, In_2O_3 [77] based gas sensors have been extensively studied because of the facile material synthesis and their high response to target gases. Among different metal oxide semiconductors, In_2O_3 is found to be an important and most promising gas-sensing material owing to its good electrical conductivity and high chemical stability. Till date, various nanostructures based on In_2O_3 for high-performance gas-sensing material in the form of thin films [78], nanowires [79], nanocrystals [80], and hollow microspheres have been developed [81].

5.3.3 Nonlinear optical properties

Nonlinear optics (NLO) is a wing of optics that explains the behaviour of light in nonlinear media, i.e. the media in which the polarization density, P responds non-linearly to the electric field, E of the light. It is noted that nanocomposite materials showed large values of optical nonlinearities and fast response time. Thus, these materials can potentially be used in areas such as image processing, optical switching, optical modulation, optical information processing, and medical applications like cancer therapy [82]. Fellahi *et al.* [83] studied the nonlinear optical properties of fluorine doped and undoped In_2O_3 thin films using X-ray diffraction, electrical resistivity, transmission and third harmonic generation. The best value of nonlinear optical susceptibility $\chi^{(3)}$ is obtained from the doped films with low electrical resistivity of 6×10^{-3} Ω cm. This is because free carrier concentration in fluorine-doped In_2O_3 samples is higher than that in undoped In_2O_3 [83].

6. Challenges and future prospect

A variety of IO/IO based nanostructured materials ranging from nanodots to nanorods, nanoneedles, nanowires or nanoplates have been obtained for various applications that are already discussed in this chapter. In case of IO based materials particularly for TCO application, the efforts have already been focused on the enhancement of electrical conductivity by adopting suitable material fabrication techniques and tuning the chemical composition with doping level of impurities. Apart from the electrical and optical properties, some other properties such as thermal stability, chemical and mechanical durability, deposition temperature, toxicity and cost of the TCO materials have also to be taken into consideration for a specific application. These properties are influenced by diverse factors, some of which being controlled by the preparation method. The main challenges for the extreme improvements of TCO as well as optoelectronics performances rely on three major areas. Firstly, it is important to understand the mechanism of structure-properties relationships and carrier mobility of TCO materials for achieving low resistivity and high transparency over extended wavelength region. Secondly, it is very much important to develop the deposition methods of IO based materials as TCO on varieties of substrates especially on temperature sensitive substrates for more efficient use of these materials and reduce the overall manufacturing costs. Moreover, it is required to perform the manufacturing and recycling techniques that would be compliant with environmental protocols. Last but not the least, fabrication/deposition of nanostructured IO based materials with reproducible properties would definitely represent a long-term opportunity in TCO industry. It is expected that the development of multicomponent IO based nanomaterials would be highly beneficial to use in various applications starting from photovoltaics to lighting, TCO, electronic devices, smart windows, gas sensors and so on.

Acknowledgements

One of the authors (HK) gratefully acknowledges CSIR, Government of India for providing financial support to his doctoral research.

Conflict of interest

The authors declare no conflict of interest.

Author details

Hasmat Khan, Saswati Sarkar, Moumita Pal, Susanta Bera and Sunirmal Jana*
Specialty Glass Division, CSIR-Central Glass and Ceramic Research Institute, Kolkata, West Bengal, India

*Address all correspondence to: sjana@cgcri.res.in; janasunirmal@hotmail.com

References

[1] Pomerantseva E, Bonaccorso F, Feng X, Cui Y, Gogotsi Y. Energy storage: The future enabled by nanomaterials. Science. 2019;**366**:eaan8285. DOI: 10.1126/science.aan8285

[2] Bierwagen O. Indium oxide—a transparent, wide-band gap semiconductor for (opto)electronic applications. Semiconductor Science and Technology. 2015;**30**:024001. DOI: 10.1088/0268-1242/30/2/024001

[3] Wang H, Liang X, Wang J, Jiao S, Xue D. Multifunctional inorganic nanomaterials for energy applications. Nanoscale. 2020;**12**:14-42. DOI: 10.1039/C9NR07008G

[4] Bera S, Pal M, Sarkar S, Jana S. Hierarchically-structured macro with nested mesoporous zinc indium oxide conducting film. ACS Applied Materials & Interfaces. 2017;**9**:4420-4424. DOI: 10.1021/acsami.6b13143

[5] Das N, Jana S, Biswas PK. Sponge-like tin doped indium oxide (ITO) from organic-inorganic composite novel precursor foam. Science of Advanced Materials. 2014;**6**:1-11. DOI: 10.1166/sam.2014.1709

[6] Khan H, Bera S, Sarkar S, Jana S. Fabrication, structural evaluation, optical and photoelectrochemical properties of soft lithography based 1D/2D surface patterned indium titanium oxide sol-gel thin film. Surface & Coatings Technology. 2017;**328**:410-419. DOI: 10.1016/j.surfcoat.2017.09.007

[7] Forsh E, Abakumov A, Zaytsev V, Konstantinova E, Forsh P, Rumyantseva M, et al. Optical and photoelectrical properties of nanocrystalline indium oxide with small grains. Thin Solid Films. 2015;**595**:25-31. DOI: 10.1016/j.tsf.2015.10.053

[8] Li P, Fan H. Porous In_2O_3 microstructures: Hydrothermal synthesis and enhanced Cl_2 sensing performance. Materials Science in Semiconductor Processing. 2015;**29**:83-89. DOI: 10.1016/j.mssp.2013.09.026

[9] Cho S. Structural, optical, and electrical properties of rf-sputtered indium oxide thin films. Journal of the Korean Physical Society. 2012;**60**:2058-2062. DOI: 10.3938/jkps.60.2058

[10] Lv Y, Li Y, Zhou R-h, Y-p P, Yao H-C, Li Z. N-Doped graphene quantum dots-decorated three-dimensional ordered macroporous In_2O_3 for NO_2 sensing at low temperatures. ACS Applied Materials & Interfaces. 2020;**12**:34245-34253. DOI: 10.1021/acsami.0c03369

[11] Cao S, Wu Y, Hou J, Zhang B, Li Z, Nie X, et al. 3D Porous Pyramid Heterostructure Array Realizing Efficient Photo-Electrochemical Performance. Advanced Energy Materials. 2019:1902935. DOI: 10.1002/aenm.201902935

[12] Haldar A, Bera S, Jana S, Bhattacharya K, Chakraborty R. Development of a cost effective surface-patterned transparent conductive coating as top-contact of LED. Journal of Applied Physics. 2014;**115**:193108. DOI: 10.1063/1.4876737

[13] Bera S, Haldar A, Pal M, Sarkar S, Chakraborty R, Jana S. Zinc-Indium-Oxide Sol-Gel Thin Film: Surface Patterning, Morphology and Photocatalytic Activity. Surface Engineering. 2015;**31**:492-501. DOI: 10.1179/1743294414Y.0000000414

[14] Groth R. Investigations on semiconducting indium oxide layers. physica status solidi. 1966;**14**:69-75. DOI: 10.1002/pssb.19660140104

[15] Jana S, Biswas PK. Effect of Zr (IV) doping on the optical properties of sol-gel based nanostructured indium

oxide films on glass. Material Chemistry and Physics. 2009;**117**:511-516. DOI: 10.1016/j.matchemphys.2009.06.038

[16] Reddy KM, Hays J, Kundu S, Dua LK, Biswas PK, Wang C, et al. Effect of Mn doping on the structural, morphological, optical and magnetic properties of indium tin oxide films. Journal of Materials Science: Materials in Electronics. 2007;**18**:1197-1201. DOI: 10.1007/s10854-007-9277-6

[17] Kundu S, Das N, Chakraborty S, Bhattacharya D, Biswas PK. Synthesis of sol-gel based nanostructured Cr(III) doped indium tin oxide films on glass and their optical and magnetic characterizations. Optical Materials. 2013;**35**:1029-1034. DOI: 10.1016/j.optmat.2012.12.005

[18] Wang X, Liu F, Chen X, Song X, Xu G, Han Y, et al. In_2O_3 Nanoparticles Decorated ZnO Hierarchical Structures for n-Butanol Sensor. ACS Applied Nano Materials. 2020;**3**:3295-3304. DOI: 10.1021/acsanm.0c00025

[19] Jiang Y, Feurer T, Carron R, Torres-Sevilla G, Moser T, Pisoni S, et al. High-mobility In_2O_3: H electrodes for four-terminal perovskite/$CuInSe_2$ tandem solar cells. ACS Nano. 2020;**14**:7502-7512. DOI: 10.1021/acsnano.0c03265

[20] Kirmani AR, Eisner F, Mansour AE, Firdaus Y, Chaturvedi N, Seitkhan A, et al. Colloidal quantum dot photovoltaics using ultra-thin, solution-processed bilayer In_2O_3/ZnO electron transport layers with improved stability. ACS Applied Energy Materials. 2020. DOI: 10.1021/acsaem.0c00831

[21] Abda JA, Mohammed WM. Indium Oxide: Synthesis and Characterization for Future Industrial Applications. AIP Conference Proceedings. 2019;**2123**: 020010. DOI: 10.1063/1.5116937

[22] Kulkarni SC, Patil DS. Synthesis and characterization of uniform spherical shape nanoparticles of indium oxide. Journal of Materials Science: Materials in Electronics. 2016;**27**:3731-3735. DOI: 10.1007/s10854-015-4215-5

[23] Luo Y, Zhang Y, Huang J. A hierarchically structured anatase-titania/indium-tin-oxide nanocomposite as an anodic material for lithium-ion battery. CrystEngComm. 2017. DOI: 10.1039/C7CE00903H

[24] Tiwari JN, Tiwari RN, Kim KS. Zero-dimensional, one-dimensional, two dimensional and three-dimensional nanostructured materials for advanced electrochemical energy devices. Progress in Materials Science. 2012;**57**:724-803. DOI: 10.1016/j.pmatsci.2011.08.003

[25] Lyakishev NP, Alymov MI, Dobatkin SV. Structural Bulk Nanomaterials. Russian Metallurgy (Metally). 2003;**3**:191-202

[26] Wei Y, Pan G, Yi H. Construction of $CdIn_2O_4$/In_2O_3 composites containing n-n heterojunctions with excellent nitro-alkanes sensing performance. Materials Today Communications. 2020;**25**:101405. DOI: 10.1016/j.mtcomm.2020.101405

[27] Shruthi J, Jayababu N, Ghosal P, Reddy MVR. Ultrasensitive sensor based on Y_2O_3-In_2O_3 nanocomposites for the detection of methanol at room temperature. Ceramics International. 2019;**45**:21497-21504. DOI: 10.1016/j.ceramint.2019.07.141

[28] Ashraf MA, Li C, Zhang D, Fakhri A. Graphene oxides as support for the synthesis of nickel sulfide–indium oxide nanocomposites for photocatalytic antibacterial and antioxidant performances. Applied Organometallic Chemistry. 2020;**34**:e5354. DOI: 10.1002/aoc.5354

[29] Haiduk YS, Khort AA, Lapchuk NM, Savitsky AA. Study of WO_3–In_2O_3 nanocomposites for highly sensitive CO and NO_2 gas sensors. Journal of Solid

State Chemistry. 2019;**273**:25-31. DOI: 10.1016/j.jssc.2019.02.023

[30] McCue JT, Ying JY. SnO_2-In_2O_3 Nanocomposites as Semiconductor Gas Sensors for CO and NO_x Detection. Chemistry of Materials. 2007;**19**:1009-1015. DOI: 10.1021/cm0617283

[31] Das N, Biswas PK. Synthesis and Characterization of Sol–Gel Foam Based Porous Mn(II) Doped Indium Tin Oxide (ITO). Journal of Nanoengineering and Nanomanufacturing. 2013;**3**:220-228. DOI: 10.1166/jnan.2013.1136

[32] Das N, Jana S, Biswas PK. Sponge-Like Tin Doped Indium Oxide (ITO) from Organic-Inorganic Composite Novel Precursor Foam. Science of Advanced Materials. 2014;**6**:252-262. DOI: 10.1166/sam.2014.1709

[33] Ismail RA. Preparation of colloidal In_2O_3 nanoparticles using nanosecond laser ablation in water. Micro & Nano Letters. 2011;**6**:951-954. DOI: 10.1049/mnl.2011.0459

[34] Yahiaa A, Attafa A, Saidia H, Dahnouna M, Khelifia C, Bouhdjera A, et al. Structural, optical, morphological and electrical properties of indium oxide thin films prepared by sol gel spin coating process. Surfaces and Interfaces. 2019;**14**:158-165. DOI: 10.1016/j.surfin.2018.12.012

[35] Ruzgar S, Caglar Y, Caglar M. The influence of low indium composition ratio on sol–gel solution-deposited amorphous zinc oxide thin film transistors. J Mater Sci: Mater Electron. 2020;**31**:11720-11728. DOI: 10.1007/s10854-020-03723-x

[36] Lee J-Y, Lee SS, Oh S-W, Lee HH, Seo Y-H, Ryua B-H, et al. Chemically improved high performance printed indium gallium zinc oxide thin-film transistors. J. Mater. Chem. 2011;**21**:17066. DOI: 10.1039/c1jm13767k

[37] Street RA, Ng TN, Lujan RA. Sol−Gel Solution-Deposited InGaZnO Thin Film Transistors. ACS Appl. Mater. Interfaces. 2014;**6**:4428-4437. DOI: 10.1021/am500126b

[38] Ghosh SS, Biswas PK, Neogi S. Effect of solar radiation at various incident angles on transparent conducting antimony doped indium oxide (IAO) film developed by sol–gel method on glass substrate as heat absorbing window glass fenestration. Solar Energy. 2014;**109**:54-60. DOI: 10.1016/j.solener.2014.08.020

[39] Hossain MA, Tianliang Z, Keat LK, Xianglin L, Prabhakar RR, Batabyal SK, et al. Synthesis of Cu(In,Ga)$(S,Se)_2$ thin films using an aqueous spray-pyrolysis approach, and their solar cell efficiency of 10.5%. J. Mater. Chem. A. 2015;**3**:4147-4154. DOI: 10.1039/C4TA05783J

[40] Lee JS, Choi W-S. Low-temperature Processing of Inkjet-printed IZO Thin-film Transistor. J Korean Phys Soc. 2014;**64**:701-705. DOI: 10.1016/j.mseb.2011.05.053

[41] Horprathum M, Eiamchai P, Kaewkhao J, Chananonnawathorn C, Patthanasettakul V, Limwichean S, Nuntawong N, Chindaudom P. Fabrication of Nanostructure by Physical Vapor Deposition with Glancing Angle Deposition Technique and its Applications. AIP Conference Proceedings 20141:617:7. DOI: 10.1063/1.4897091

[42] Gwamuri J, Marikkannan M, Mayandi J, Bowen PK, Pearce JM. Influence of Oxygen Concentration on the Performance of Ultra-Thin RF Magnetron Sputter Deposited Indium Tin Oxide Films as a Top Electrode for Photovoltaic Devices. Materials. 2016;**9**:63. DOI: 10.3390/ma9010063

[43] Chang TH, Chang SJ, Chiu CJ, Wei CY, Juan YM, Weng WY. Bandgap-Engineered in Indium-Gallium-Oxide Ultraviolet Phototransistors. IEEE

Photon. Technol. Lett. 2015;**27**:915-918. DOI. 10.1109/LPT.2015.2400116

[44] Ma Q, Zheng H-M, Shao Y, Zhu B, Liu W-J, Ding S-J, et al. Atomic-Layer-Deposition of Indium Oxide Nano-films for Thin-Film Transistors. Nanoscale Res. Lett. 2018;**13**(4). DOI: 10.1186/s11671-017-2414-0

[45] Cindemir U, Lansåker PC, Österlund L, Niklasson GA, Granqvist C-G. Sputter-Deposited Indium–Tin Oxide Thin Films for Acetaldehyde Gas Sensing. Coatings. 2016;**6**:19. DOI: 10.3390/coatings6020019

[46] Khan GG, Ghosh S, Sarkar A, Mandal G, Mukherjee GD, Manju U, et al. Defect engineered d 0 ferromagnetism in tin-doped indium oxide nanostructures and nanocrystalline thin-films. Journal of Applied Physics. 2015;**118**:074303. DOI: 10.1063/1.4928952

[47] Yang C-S, Chang C-M, Chen P-H, Yu P, Pan C-L. Broadband terahertz conductivity and optical transmission of indium-tin-oxide (ITO) nanomaterials. OPTICS EXPRESS. 2013;**21**:16670-16682. DOI: 10.1364/OE.21.016670

[48] Y-g K, Tak YJ, Kim HJ, Kim W-G, Yoo H, Kim HJ. Facile fabrication of wire-type indium gallium zinc oxide thin film transistors applicable to ultrasensitive flexible sensors. Scientific Reports. *2018*;*8*:5546. DOI: 10.1038/s41598-018-23892-4

[49] Haldar A, Bera S, Jana S, Bhattacharya K, Chakraborty R. Development of a cost effective surface-patterned transparent conductive coating as top-contact of light emitting diodes. Journal of Applied Physics. 2014;**115**:193108. DOI: 10.1063/1.4876737

[50] Leung S-F, Zhang Q, Xiu F, Yu D, Ho JC, Li D, Fan Z. Light Management with Nanostructures for Optoelectronic Devices. J. Phys. Chem. Lett. 2014; 5:1179 1195. Doi.org/10.1021/jz500306f

[51] Zhao X-M, Xia Y, Whitesides GM. Soft lithographic methods for nano-fabrication. J. Mater. Chem. 1997;7:1069-1074

[52] Liu J, Wöll C. Surface-supported metal–organic framework thin films: fabrication methods, applications, and challenges. Chem. Soc. Rev. 2017;**46**:5730-5770. DOI: 10.1039/c7cs00315c

[53] Bera S, Lee SA, Kim C-M, Khan H, Jang HW, Kwon S-H. Controlled synthesis of vertically aligned SnO_2 nanograss-structured thin films for $SnO_2/BiVO_4$ core–shell heterostructures with highly enhanced photoelectrochemical properties. Chemistry of Materials. 2018;**30**:8501-8509

[54] Suzuki R, Nishi Y, Matsubara M, Muramatsu A, Kanie K. Single-Crystalline Protrusion-Rich Indium Tin Oxide Nanoparticles with Colloidal Stability in Water for Use in Sustainable Coatings. ACS Appl. Nano Mater. 2020;**3**:4870-–4879 https://dx.doi.org/10.1021/acsanm.0c01023

[55] Frei MS, Mondelli C, Cesarini A, Krumeich F, Hauert R, Stewart JA, et al. Role of Zirconia in Indium Oxide-Catalyzed CO2 Hydrogenation to Methanol. ACS Catal. 2020;**10**:1133-1145. DOI: 10.1021/acscatal.9b03305

[56] Shah S, Hussain S, Qiao G, Tan J, Javed MS, Zulfiqar GC, et al. Decorating spherical In_2O_3 nanoparticles onto ZnO nanosheets for outstanding gas-sensing performances. Journal of Materials Science: Materials in Electronics. 2020;**31**:3924-3933 https://doi.org/10.1007/s10854-020-02937-3

[57] Jothibas M, Manoharan C, Jeyakumar SJ, Praveen P. Study on

structural and optical behaviors of In_2O_3 nanocrystals as potential candidate for optoelectronic devices. J Mater Sci: Mater Electron. 2015. DOI: 10.1007/s10854-015-3623-x

[58] Bykova LE, Zharkov SM, Myagkov VG, Zhigalov VS, Patrin GS. In Situ Electron Diffraction Investigation of Solid State Synthesis of Co-In_2O_3 Ferromagnetic Nanocomposite Thin Films. JOM. 2020;**72**:2139-2145 https://doi.org/10.1007/s11837-019-03919-5

[59] Babu SH, Kaleemulla S, Rao NM, Krishnamoorthi C. Indium oxide: A transparent, conducting ferromagnetic semiconductor for spintronic applications. Journal of Magnetism and Magnetic Materials. DOI. http://dx.doi.org/10.1016/j.jmmm.2016.05.007

[60] Mukherji R, Mathur V, Samariya A, Mukherji M. Experimental and theoretical assessment of Fe-doped indium-oxide-based dilute magnetic semiconductors. Philosophical Magazine. 2019 https://doi.org/10.1080/14786435.2019.1624851

[61] Ares MCZ, Pérez MR, Carballo LQ, Fernández JM. Assessment of clays fromPuertollano (Spain) for their use in fine ceramic by diffuse reflectance spectroscopy. Applied Clay Science. 2015;**108**:135-143. DOI: doi.org/10.1016/j.clay.2015.02.010

[62] Swallow JEN, Williamson BAD, Sathasivam S, Birkett M, Featherstone TJ, Murgatroyd PAE, et al. Resonant doping for high mobility transparent conductors: the case of Mo-doped In_2O_3. Mater.Horiz. 2020;**7**:236-243

[63] Qiao F, Lu L, Han P, Ge D, Rui Y, Gu D, et al. A Combined Experimental and Theoretical Study of Screen-printing High Transparent C onductiveMesoscopic ITO Films. Scientific Reports. 2020;**10**:5024 https://doi.org/10.1038/s41598-020-61124-w

[64] Afre RA, Sharma N, Sharon M, Sharon M. Transparent conducting oxide films for various applications: a review. Rev.Adv.Mater.Sci. 2018;**53**:79-89

[65] Simya OK, Balachander K, Dhanalakshmi D, Ashok AM. Performance of different anti-reflection coating and TCO layers for kesterite based thin film photovoltaic devices using Essential Macleod simulation program. Superlattices Microstruct. 2020;**145**:106579. DOI: 10.1016/j.spmi.2020.106579

[66] Park H-W, Ji S, Lim H, D-w C, Park J-S, Chung K-B. Anti-reflective conducting indium oxide layer on nanostructured substrate as a function of aspect ratio. Appl. Phys. Lett. 2016;**109**:121604. DOI: 10.1063/1.4963265

[67] Khan I, Bauch M, Dimopoulos T, Dostalek J. Nanostructured as-deposited indium tin oxide thin films for broadband antireflection and light trapping. Nanotechnology. 2017;**28**:325201. DOI: 10.1088/1361-6528/aa79df

[68] Llordés A, Garcia G, Gazquez J, Milliron DJ. Tunable near-infrared and visible-light transmittance in nanocrystal-in-glass composites. Nature. 2013;**500**:323-326. DOI: 10.1038/nature12398

[69] Mashkov O, Körfer J, Eigen A, Yousefi-Amin A-A, Killilea N, Barabash A, et al. Effect of Ligand Treatment on the Tuning of Infrared Plasmonic Indium Tin Oxide Nanocrystal Electrochromic Devices. Advanced Engineering Materials. 2020;**22**:2000112. DOI: 10.1002/adem.202000112

[70] Luceño-Sánchez JA, Díez-Pascual AM, Capilla RP. Materials for Photovoltaics: State of Art and Recent Developments. International Journal of Molecular Sciences. 2019;**20**:976. DOI: 10.3390/ijms20040976

[71] Polino G, Dell'Elce S, Liscio A, Notte LL, Cardone G, Carlo AD, et al. Indium Tin Oxide–Based Fully Spray-Coated Inverted Solar Cells with Nontoxic Solvents: The Role of Buffer Layer Interface on Low-Bandgap Photoactive Layer Performance. Energy Technology. 2019;7:1800627. DOI: 10.1002/ente.201800627

[72] Kang JH, Song A, Park YJ, Seo JH, Walker B, Chung K-B. Tungsten-Doped Zinc Oxide and Indium–Zinc Oxide Films as High-Performance Electron-Transport Layers in N–I–P Perovskite Solar Cells. Polymers. 2020;**12**:737. DOI: 10.3390/polym12040737

[73] Lee MG, Park JS, Jang HW. Solution-Processed Metal Oxide Thin Film Nanostructures for Water Splitting Photoelectrodes: A Review. Journal of the Korean Ceramic Society. 2018;**55**:185-202. DOI: 10.4191/kcers.2018.55.3.08

[74] Chen K-Y, Chang S-P, Lin C-h. Effect of oxygen vacancy concentration on indium tungsten oxide UV-A photodetector. RSC Advances. 2019;**9**:87-90. DOI: 10.1039/C8RA08803A

[75] Huang Y, Li Y, Deng R, Yao B, Ding Z, Zhang L, et al. Highly spectrum-selective near-band-edge ultraviolet photodiode based on indium oxide with dipole-forbidden bandgap transition. Ceramics International. 2016;**42**:8017-8021. DOI: 10.1016/j.ceramint.2016.01.206

[76] Meng M, Wu X, Zhu X, Zhu X, Chu PK. Facet Cutting and Hydrogenation of In2O3 Nanowires for Enhanced Photoelectrochemical Water Splitting. ACS Appl. Mater. Interfaces. 2014;**6**:4081-4088. DOI: 10.1021/am4056358

[77] Chava RK, Cho HY, Yoon JM, Yu YT. Fabrication of aggregated In_2O_3 nanospheres for highly sensitive acetaldehyde gas sensors. Journal of Alloys and Compounds. 2019;**772**:834-842. DOI: 10.1016/j.jallcom.2018.09.183

[78] Carlos-Chilo AF, Luza-Mamani LG, Baldárrago-Alcántara AA, Aragón FFH, Pacheco-Salazar DG. Study of columnar growth polycrystalline (Sn, Cr) co-doped In_2O_3 films deposited by sputtering technique for potential gas sensors applications. Vacuum. 2018;**157**:475-480. DOI: 10.1016/j.vacuum.2018.08.032

[79] Park S. Acetone gas detection using TiO_2 nanoparticles functionalized In_2O_3 nanowires for diagnosis of diabetes. Journal of Alloys and Compounds. 2017;**696**:655-662. DOI: 10.1016/j.jallcom.2016.11.298

[80] Anand K, Kaur J, Singh RC, Thangaraj R. Preparation and characterization of Ag-doped In_2O_3 nanoparticles gas sensor. Chemical Physics Letters. 2017;**682**:140-146. DOI: 10.1016/j.cplett.2017.06.008

[81] Hu J, Sun YJ, Xue Y, Zhang M, Chen Y. Highly sensitive and ultra-fast gas sensor based on CeO_2-loaded In_2O_3 hollow spheres for ppb-level hydrogen detection. Sensors and Actuators B Chemical. 2018;**257**:124-135. DOI: 10.1016/j.snb.2017.10.139

[82] Zhong J, Xiang W. Influence of In_2O_3 nanocrystals incorporation on sodium borosilicate glass and their nonlinear optical properties. Materials Letters. 2017;**193**:22-25. DOI: 10.1016/j.matlet.2017.01.062

[83] Fellahi N, Addou M, Sofiani Z, Jouad ME, Bahedi K, Bayoud S, et al. Effect of fluorine incorporation on the third-order nonlinear-optical properties of indium oxide thin films prepared by spray pyrolysis. Journal of Optoelectronics and Advanced Materials. 2010;**12**:1087-1091

Chapter 2

Indium Chalcogenide Nanomaterials in the Forefront of Recent Technological Advancements

Siphamandla C. Masikane and Neerish Revaprasadu

Abstract

In the last decade, there has been an increasing trend in the exploitation of indium chalcogenides in various applications which range from water splitting reactions in renewable energy to degradation of dyes in environmental rehabilitation. This trend is attributed to the interesting and unique properties of indium chalcogenide nanomaterials which can be easily tuned through a common approach: particle size, shape and morphology engineering. In this chapter, we outline the preferred attributes of indium chalcogenide nanomaterials which are deemed suitable for recent applications. Furthermore, we explore recent reaction protocols which have been reported to yield good quality indium chalcogenide nanomaterials of multinary configurations, e.g. binary and ternary compounds, among others.

Keywords: sulfide, selenide, telluride, multinary, applications

1. Introduction

Over the years, there has been an increasing demand on state-of-art solutions to solve real world problems such as the energy crises and efficient early-detection protocols in biomedical services. The current systems in place suffer from a range of issues, e.g. an increase in the depletion rate of fossil fuel and petroleum reserves as precursors in the electrical power generation plants [1]. Although, in the context of electricity generation, there exists alternatives such as nuclear power, unwavering challenges such as toxicity of nuclear waste still persist [2]. Another good example is the use of conventional dyes for the detection of tumors (typically having issues with stability and sensitivity) and drug delivery systems, which both generally lack selectivity i.e. in crucial need of smart, guide-assisted delivery to an affected target area [3]. As a response to these issues, among many that exist, scientists and engineers have presented a range of nanotechnology-based solutions through successes in the development and pioneering work on functional nanomaterials and related devices. There are, however, reservations in trusting these technologies in the general public domains, attributed to insufficient knowledge and/or lack of educational strategies [4]. Thus, progress in introducing these systems for general use still remains a challenge, with few successes such as QLED televisions [5] already available to the general public consumers for everyday use.

The core fundamental principle to grasp on nanotechnology and nanomaterials is that when the particle size dimensions of a bulk material decrease to the nanometer scale, improved and/or novel properties emerge. Thus, properties of a material can be tuned to desired standards best suited for specific applications, by simply manipulating particle size and shape. Indium chalcogenide nanomaterials are among many functional materials which boast rich literature in the aforementioned context, hence, their technological importance continues to be showcased in widespread applications to date.

The surge in the interest of indium chalcogenide nanomaterials has mainly been fueled by their recognition as alternative candidates against giants in the field of photovoltaics and sustainable energy solutions, such as cadmium chalcogenides which are known for their toxicity issues albeit achieving high performance and efficiency in metal chalcogenide-based semiconductor solar cells and other optoelectronic applications [6]. There are other less-to-non-toxic candidates which have been identified, such as antimony [7] and tin [8] chalcogenides, among others. However, indium chalcogenides contain a broad spectrum of crystallographic phases/species which exhibit unique properties attributed to different atomic compositions and crystal lattice orientation (polymorphism), contrary to antimony and tin chalcogenides. An example of this can be seen in the indium sulfide series, where InS, In_3S_4, In_6S_7 and In_2S_3 (α-In_2S_3, β-In_2S_3 and γ-In_2S_3 [9]) phases have been obtained experimentally [10]. This, in addition to manipulating particle size and shape, as well as employing other enhancement techniques such as doping and composite fabrications, present endless opportunities to harness tailor-made properties.

The most common and easy route to tune the properties of nanomaterials is by tweaking reaction parameters during synthesis. Therefore, the choice of synthetic methods best suited for specific precursors is of crucial importance [11]. Over the years, there has been an intensive research invested on precursor design necessary to produce high quality nanomaterials [12]. Hence, metalorganic compounds gained unprecedented attention as molecular precursors compatible with a range of fabrication protocols. These molecular precursors have made it possible to access various classes of nanomaterials, although the overall nanomaterial fabrication protocols were initially a hit-or-miss process. As a result of this approach, useful data has been obtained which has formed an integral part of theoretical models used to predict novel nanomaterials and their corresponding properties. As much as molecular precursors have demonstrated their preference and superiority over conventional salt-based precursors in the context of nanomaterial fabrication, the latter is however currently ideal for the development of devices which are sensitive to impurities, among other factors. Hence, recent technological advances (from late 2019 to date of this book chapter) of indium chalcogenide nanomaterials presented in the next sections are predominantly obtained through conventional salt-based precursor routes. Interesting literature on molecular precursors for indium chalcogenide nanomaterials is available elsewhere [13, 14].

2. Indium sulfide series

Research on indium chalcogenide nanomaterials predominantly focuses on the indium sulfide series, attributed to readily available, abundant, cheap and stable precursors. This series finds applications in various applications, typically in optoelectronics. Among recent advancements is the selective NO_2 gas sensing abilities of β-In_2S_3 thin films prepared by spray pyrolysis; this preliminary study introduces β-In_2S_3 thin films as less toxic and cheaper alternatives to highly selective and sensitive cadmium sulfide-based NO_2 gas sensors [15]. In other work, In_2S_3

thin films prepared by ultrasonic spray pyrolysis was evaluated, for the first time, as photoelectrodes (**Figure 1**) for all-vanadium photoelectrochemical batteries [16]. The efficiency was linked to the degree of optical and photoelectrochemical behavior associated with the thickness of the In_2S_3 thin films. In both works, it becomes apparent that the physical alterations of the films are necessary to improve selectivity, sensitivity and efficiency. There are two notable recent reports which have provided preliminary solutions as per above: (i) band gap (1.9–2.3 eV) and electrical resistivity (5.5×10^0–6.0×10^3 Ωm) control through thermal treatment of as-prepared In_2S_3 thin films at different temperatures in the presence/absence of sulfur powder [17], and (ii) tunable morphological (root mean square roughness) and optical properties (transmittance and photoluminescence) of the In_2S_3 thin films by varying the S/In molar ratio in spray pyrolysis deposition experiments [18].

Similar to the research objectives in Ref [15], indium sulfide is yet again demonstrated as a promising alternative to the cadmium sulfide, in this case as a buffer layer in the $Cu(In,Ga)Se_2$ solar cell [19]. It was found that the indium sulfide-based solar cell achieved 15.3% efficiency compared to 17.1% recorded for the cadmium sulfide counterpart. The authors report that the observed efficiency is attributed to substrate temperature optimization during the sputtering method-based experiments. According to the study, the increase in substrate temperature tempers with the In_xS_y and $Cu(In,Ga)Se_2$ compositions; an increase in temperature resulted to a sulfur-rich In_xS_y buffer layer, as well as copper depletion observed in the $Cu(In,Ga)Se_2$ absorber layer, as seen in **Figure 2**. Furthermore, sodium doping was observed in both the In_xS_y layer and in the In_xS_y and $Cu(In,Ga)Se_2$ interface. Thus, it is these features which were identified to play a major role in the increase of the solar efficiency.

Other efforts to improve attractive properties of In_2S_3 thin films have been reported, such as silver doping as means of improving electrical transport [20], as well as plasma treatment which consequently results to the self-formation of metallic indium arrays at the surface thus presenting opportunities in fabricating

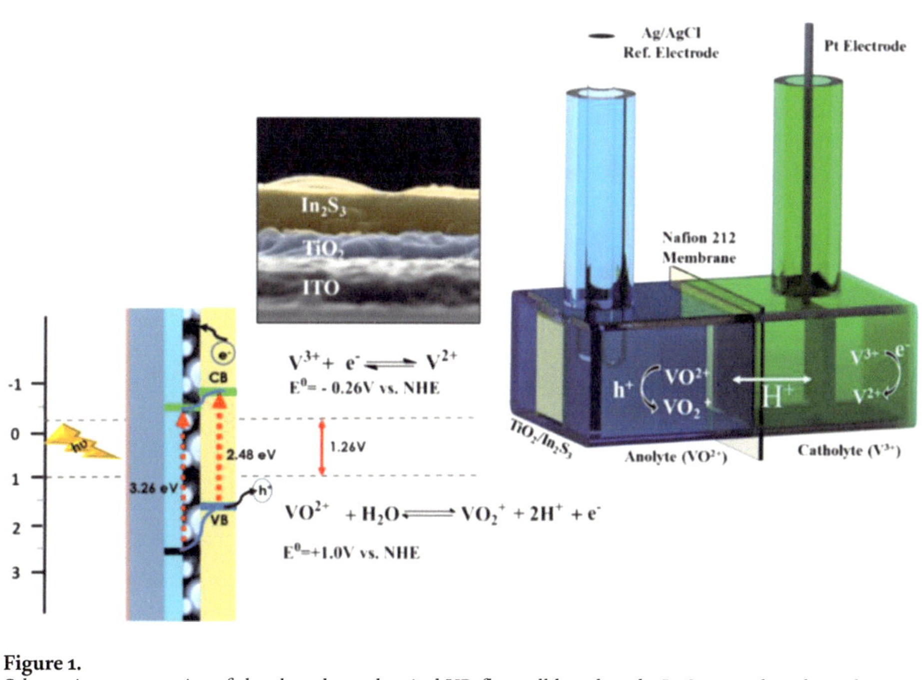

Figure 1.
Schematic representation of the photoelectrochemical VR-flow cell based on the In_2S_3-type photoelectrode. Reprinted with permission from Ref. [16]. Copyright 2020 American Chemical Society.

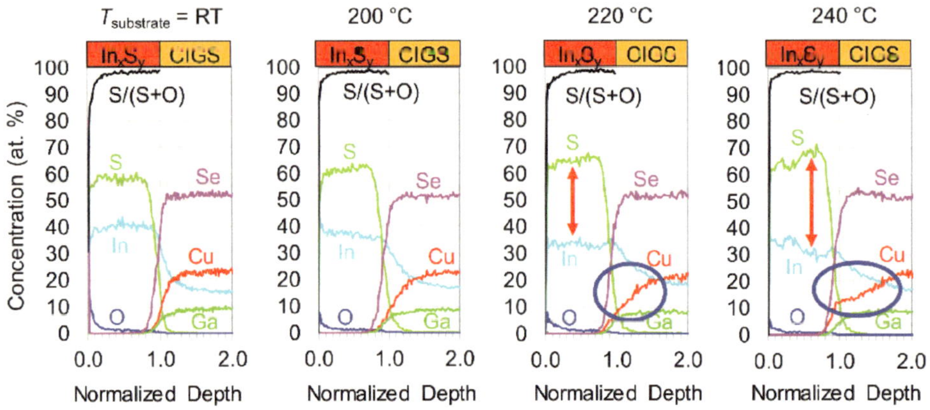

Figure 2.
Elemental composition of In_xS_y and $Cu(In,Ga)Se_2$ layers deposited at different substrate temperatures. Reprinted with permission from Ref. [19]. Copyright 2020 MDPI.

heterostructures for potential use optoelectronics [21]. Bilayer and trilayer InS triangular nanoflakes have also been prepared by chemical vapor deposition [22], potential applications envisaged as heterojunctions in nanoelectronic devices.

The films outlined above are predominantly obtained from existing technologies such as spray pyrolysis, thermal evaporation and chemical vapor deposition, where the films are directly prepared on a substrate. A new, solution-based synthesis of suspended 2D ultrathin sheets was developed [23]. This novel strategy, optimized through the synthesis of In_2S_3 sheets, exploits a self-assembling anisotropic growth mechanism templated by a combination of amine ligand with a geometrically-matched alkane. The obtained In_2S_3 sheets exhibited high photoelectric activities best suited for photoelectrochemical applications. Preliminary experiments displayed versatility of the method, attributed to the successful preparation of other 2D nanostructures such as Co_9S_8, MnS, SnS_2, Al_2S_3 and MoS_2. Thus, this presents an alternative route to easily prepare functional thin films which could ultimately be transferred to desired substrate post preparation and manipulation processes.

It has been observed that the recent advances in indium sulfide nanomaterials outlined above predominantly use the multiple precursor route. Although progress has been made in the past few years, the search for novel metalorganic single-source precursors for indium sulfide continues. New indium complexes with aminothiolate ligands have been synthesized and characterized fully [24], their structures are provided in **Figure 3**. Preliminary evaluations as potential single-source precursors showed that complex **1** is able produce β-In_2S_3 nanoparticles, complexes **2** and **3**

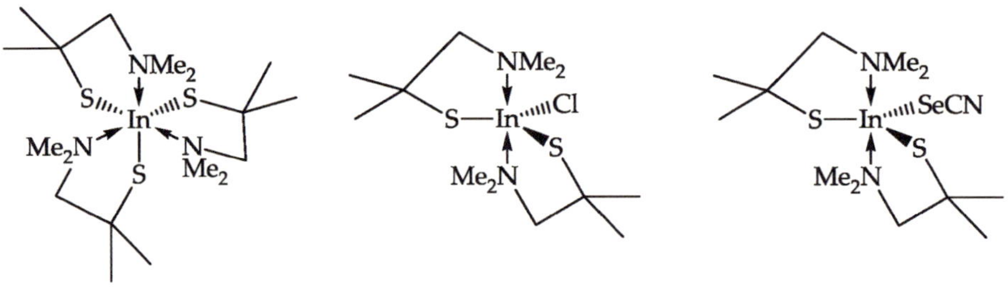

1 **2** **3**

Figure 3.
Chemical structures of novel indium (III) aminothiolate complexes prepared by Ref. [24].

however need extensive work, as the diffraction studies for phase identification were inconclusive. However, microelemental analyses suggest that the nanomaterial exhibit general formulae In_2S_3 and In_2Se_2S from complexes **2** and **3**, respectively.

2.1 Indium sulfide-based ternary and quaternary nanomaterials

The main interest on indium sulfide-based ternary and quaternary nanomaterials is their optical properties, predominantly exploited in emission/photoluminescence applications. Recent studies in this field has focused on the chalcopyrite-type materials, copper indium sulfide ($CuInS_2$) and silver indium sulfide ($AgInS_2$) in particular. A recent, concise review on the synthesis and applications of $CuInS_2$ is available in Ref [25]. However, there are interesting literature reports which emerged subsequent to the publication of the review. For example, there is a study which has evaluated the influence of halide ions on the optical properties of $CuInS_2$ quantum dots [26]. Similar to our work where we evaluated the influence of halide ligands in the single-source precursors on the morphological and optical properties of cadmium sulfide [27, 28] and lead sulfide [29] nanoparticles, the authors in this case follow a multiple-source precursor route (through the solvothermal synthetic protocols) using CuX (where X = I, Cl and Br) salts. The optical properties show unique behavior with respect to the metal salt used, attributed to the physicochemical properties resulting from the growth processes which consequently promote accumulation of the halide ions in the crystal lattices of the quantum dots. In another report, the importance of controlling the Cu:In ratio

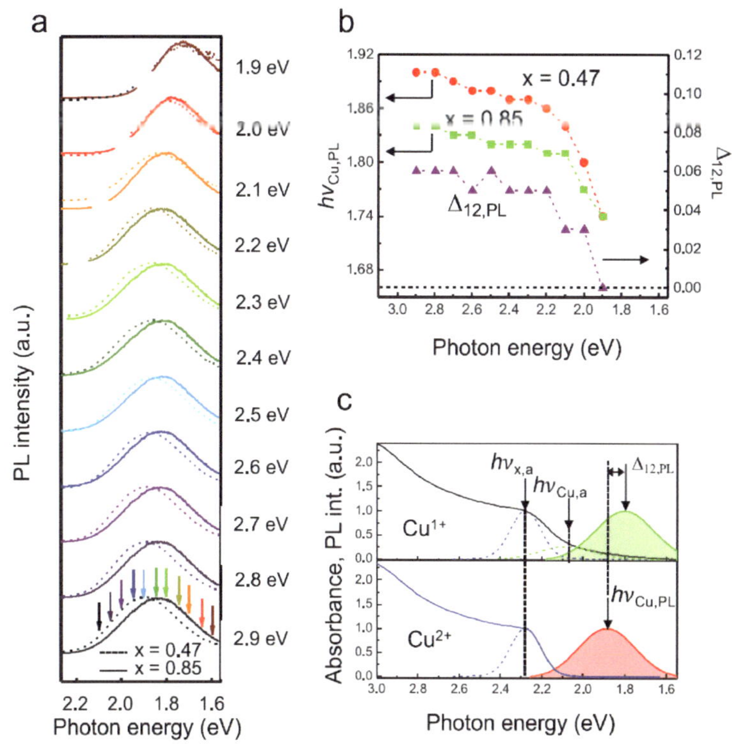

Figure 4.
(a) Resonant photoluminescence (PL) measurements of Cu_xInS_2 quantum dots where x = 0.47 (dashed lines) and x = 0.85 (solid lines), at different excitation energies. (b) the PL peak energies extracted from (a). (c) Simulated absorption (lines) and corresponding PL (shaded peaks) spectra of Cu_xInS_2 quantum dots with respect to Cu^{1+} and Cu^{2+} defects. Reprinted with permission from Ref. [30]. Copyright 2020 American Chemical Society.

in $CuInS_2$ quantum dots to harness different properties for various applications is discussed [30]. The report suggests that the change in the Cu:In ratio induces defects attributed to what the authors refer to as Cu^{1+} *vs* Cu^{2+} concentration defects, resulting in different optical emission behaviors as observed in **Figure 4**.

Although $CuInS_2$ is a reputably known non-toxic material displaying attractive properties which are already exploited intensively in biomedical-based applications, there are however recent reports which have shown compelling experimental evidence contradicting this non-toxic behavior. A recent research study has observed the instability of zinc sulfide (ZnS) shell-free $CuInS_2$ quantum dots relative to the shelled counterparts in the *in vitro* studies [31], degradation was demonstrated by rapid dissolution in simulated biological fluid (SBF) and artificial lysosomal fluid (ALS) through absorption spectroscopy measurements shown in **Figure 5**. Furthermore, it was demonstrated that shell-free $CuInS_2$ induces severe toxicity in the *in vivo* studies compared to the infamous, toxic cadmium selenide. In another report, $CuInS_2$ nanocrystals were exposed in environment-like conditions (including alkaline and acidic settings) thereby promoting weathering [32]. It was observed that when the environmental pathogenic bacteria *Staphylococcus aureus* CMCC 26003 strain is exposed to weathered $CuInS_2$ nanocrystals, it develops increased tolerance to certain antibiotics such as penicillin G, tetracycline and ciprofloxacin. Thus, these two studies are a constant reminder with regards to creating awareness that alternative "green" approaches require concise evaluations and any possible adverse effects towards disruption of natural and crucial processes.

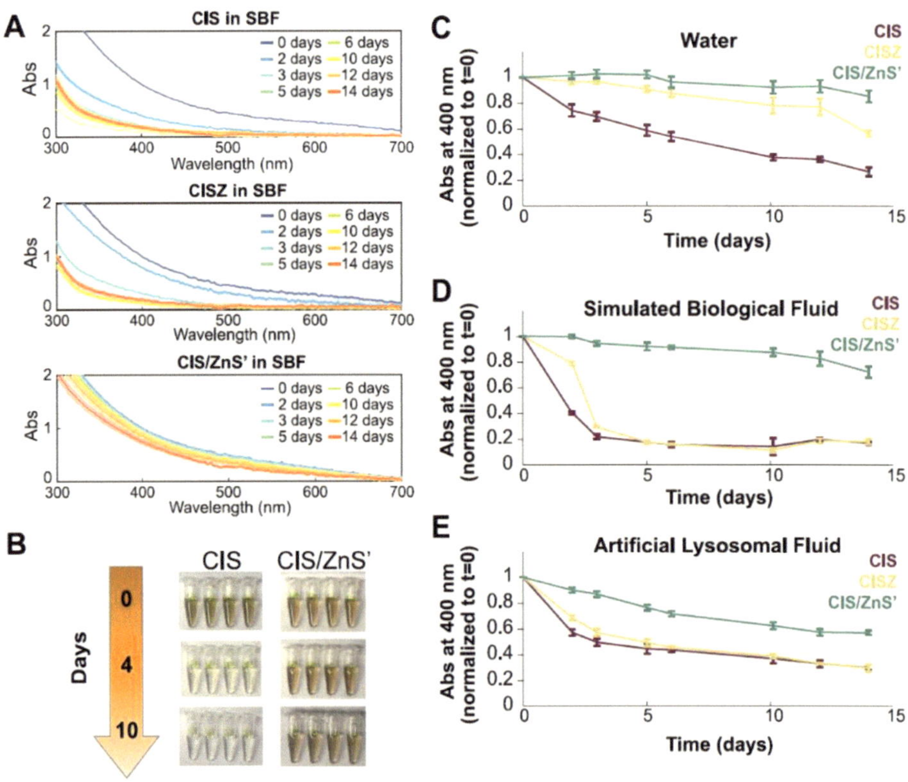

Figure 5.
(A) Comparative study on the dissolution of CIS ($CuInS_2$), CISZ (zinc-alloyed $CuInS_2$) and CIS/ZnS' ($CuInS_2$/ZnS core/shell) quantum dots by absorption spectroscopy measurements. (B) Visual evidence of dissolution in SBF. (C-E) dissolution studies in various media. Reprinted with permission from Ref. [31]. Copyright 2020 American Chemical Society.

This should however not deter attempts in developing similar technologies such as $AgInS_2$ quantum dots which have recently shown ultralong PL decay time attributed to the coordinating ligands which bear electron rich groups capable of passivating surface trap centers and achieving strong emissions [33].

Apart from the chalcopyrite series, other indium sulfide-incorporated multinary nanomaterials have made significant technological progress. Recently, $CdIn_2S_4$ and $ZnIn_2S_4$ nanostructures have been prepared by solvent-free green reaction protocols at moderately low temperatures [34]; the nanostructures displayed good photocatalytic activities in hydrogen evolution reactions through the splitting of hydrogen sulfide and water under visible light conditions. The activities were however lower than those of other $CdIn_2S_4$ and $ZnIn_2S_4$ nanostructures reported elsewhere [35, 36], attributed to synthetic the method limitations particularly on poor control of physical features such as particle size and shape. The quaternary system which has been recently reported is the $Zn_{2x}Cu_{1-x}In_{1-x}S_2$ [37] and $Zn_{2x}Ag_{1-x}In_{1-x}S_2$ [38] nanomaterials which display unique optical properties with varying composition and an active component potential in the light harvesting inorganic–organic hybrid nanomaterial, respectively.

3. Indium selenide series

The indium selenide series exhibits similar characteristics to the indium sulfide series, such as multiple crystallographic phases and polymorphic materials which have unique properties already found use in various applications. The chemistry, synthesis and application of the indium selenide series is already disseminated in comprehensive literature reviews available elsewhere [39, 40].

Among recent developments in the synthesis of indium selenide, is a novel reaction protocol which has been designed to growing ultrathin films of stoichiometric indium selenide (InSe) by precipitation of the thermally evaporated InSe crystal on a chemically neutral oil [41]. In another study, the thermal evaporation technique was used albeit to synthesize InSe nanowires on silicon and quartz silica substrates through an edge-epitaxial growth mechanism [42], this work presents a solution on challenges associated with growing nanowires on these substrates as a result of the lattice mismatch. This provides easy access to investigate the efficiency of nanowires on fabricated electronic and optoelectronic devices. The epitaxial growth approach has also been employed in the fabrication of few-layer β-In_2Se_3 thin films on c-plane sapphire and silicon substrates through the metalorganic chemical vapor deposition method [43], the synthetic protocols have potential scale up capabilities while retaining good quality uniform film. Obtaining defect-free nanomaterials from bulk counterparts through exfoliation mediated processes still remains an economically ideal route, however, the most common issue is low yields. Recent efforts towards this direction is the development of ultrafast electrochemical-assisted delamination of bulk In_2Se_3 through intercalation by tetrahexylammonium ions in a typical setup provided in **Figure 6(d)** [44], the authors demonstrated that the results are reproducible and the obtained yields of up to 83% flakes which have large micron-scale lateral sizes suitable for fabricating various nanodevices.

Applications of binary indium selenide nanomaterials are provided in **Table 1**. As observed, the choice of synthetic method is crucial since it produces nanomaterials suitable for specific applications. Recent interests are towards synthesizing good quality nanosheets and thin films, attributed to the development of novel next-generation devices for use in various fields. It is apparent that the sought-after features of binary indium selenide nanomaterials are optical properties-related, hence exploitation predominantly observed in optoelectronic applications.

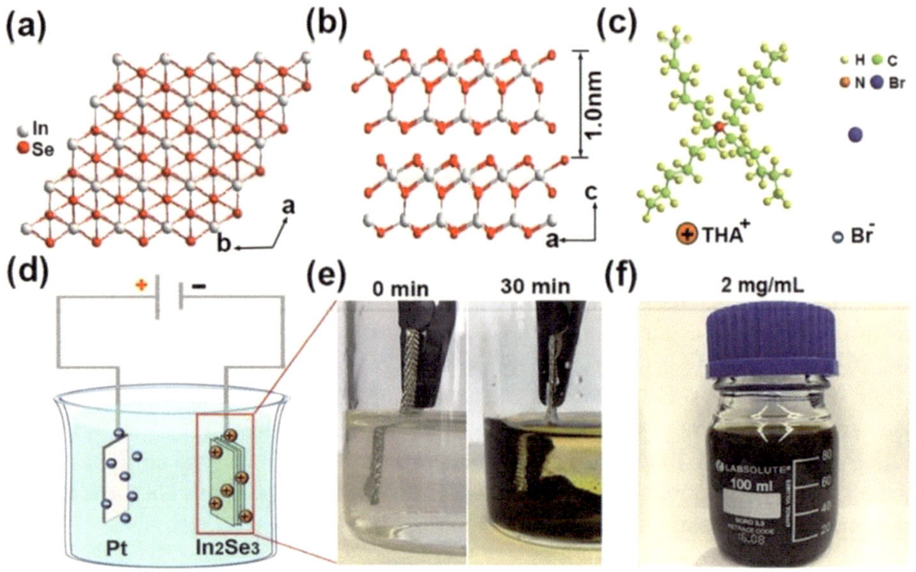

Figure 6.
(a) Top and (b) side view of the layered crystal structure In_2Se_3; (c) the chemical structure of the tetrahexylammonium-based intercalant; (d) experimental setup; (e) images showing the beginning and completion of the experiment; (f) dispersion of delaminated In_2Se_3 nanosheets in dimethylformamide. Reprinted with permission from Ref. [44]. Copyright 2020 WILEY-VCH.

Material	Synthetic method	Application	Material type	Reference
InSe	Mechanical exfoliation (Scotch tape)	Saturable absorber for mid-infrared pulsed laser	Thin films	[45]
Sn-doped InSe		Photoluminescent sensor for sulfur vapors	Nanosheets	[46]
InSe		Field-effect transistor		[47–49]
		Field-effect transistor for pressure sensors		[50]
	Edge-epitaxial growth	Photodetector	Nanowires	[42]
	Liquid phase exfoliation	All-optical diodes and switching	Nanosheets	[51]
α-In_2Se_3		Potential use in Ultrafast photonic devices		[52]
γ-In_2Se_3	Electrosynthesis	Electrocatalyst for carbon dioxide electroreduction to Syngas	Nanoparticles	[53]
α-In_2Se_3	Mechanical exfoliation (Scotch tape)	Photodetector	Nanosheets	[54]
In_2Se_3	Electrochemical-based exfoliation			[44]
In_3S_4	Electron-beam deposition	Potential applications in electrical and thermoelectrical devices	Thin films	[55]

Table 1.
Recent advances in the application of binary indium selenide nanomaterials.

3.1 Indium selenide-based ternary and quaternary nanomaterials

Multinary indium selenide-based nanomaterials, with the exception of the binary system, are rarely subjects of research interest compared to the sulfide counterparts, most probably due to synthetic challenges associated with limited economic precursors. Hence, recent technologies outlined in section 3 above rely mostly on pre-synthesized (at extreme reaction conditions) and commercial indium sulfide bulk material. Regardless of this, recent efforts on multinary indium selenide-based nanomaterials have been reported.

Silver indium sulfide nanocrystals of the $AgIn_5Se_8$ phase have been synthesized through an eco-friendly electrochemical method using L-glutathione as a stabilizing agent [56]. The photoluminescence spectra of the nanocrystals showed an increase in quantum yields with an increase in silver-to-indium ratio used during synthesis. Furthermore, the nanocrystals displayed good photothermal responses which are ideal for hyperthermia applications. Layered manganese indium sulfide nanosheets of the $MnIn_2Se_4$ phase prepared by mechanical exfoliation, have recently been demonstrated as a potential candidate for use in magnetic and optoelectronic devices due to their interesting magnetic and transport properties [57]. Computational studies using first-principle calculations have predicted properties of the layered indium selenide bromide (InSeBr) which have significantly been ignored [58]. The comprehensive Raman scattering measurements have predicted that InSeBr would be a good potential candidate for use in optoelectronic properties. Research interests on quaternary indium selenide-based nanomaterials have primarily focused on copper indium gallium selenide [$Cu(In,Ga)Se_2$] materials which are heavily invested in the fabrication of next-generation semiconductor solar cells; a recent, comprehensive review on the science, synthesis and application of $Cu(In,Ga)Se_2$ nanomaterials is available elsewhere [59].

4. Indium telluride series

Indium telluride and derived nanomaterials are rarely common, due to a limited application scope. The most common application of indium telluride nanomaterials is in thermoelectrics. There has been attempts in gas sensing applications showing unsatisfactory sensitivity, attributed to the low electrical resistance of the nanomaterial [60]. Other applications have been mentioned elsewhere with references therein [61]. In a recent report, the authors devised a method of preparing In_2Te_3 thin films composed of nanowire structures from bulk InTe using a chemical vapor deposition technique through a gold-catalyzed vapor-liquid–solid growth mechanism [62]. It was however observed that the low electrical resistivity and thermal conductivity cannot be improved by simply changing the morphology of the particles. A separate study has reported that these properties can be effectively improved by doping In_2Te_3 with aluminum and antimony [63]. The stoichiometric InTe phase is also used in thermoelectric applications; recent studies also identify that the thermoelectric performance is improved by doping with antimony [64].

4.1 Indium telluride-based ternary and quaternary nanomaterials

Ternary analogues of indium tellurides also find use in thermoelectric applications, such as copper indium telluride ($CuInTe_2$) and silver indium sulfide ($AgInTe_2$). The thermoelectric properties of the former have recently been reported to be enhanced by doping with manganese [65], while for the latter, adjusting only the silver concentration x (in $Ag_{1-x}InTe_2$) was sufficient [66]. The interesting properties of another ternary material potassium indium telluride ($KInTe_2$), for the first time, have been recently predicted and investigated through theoretical first-principle calculations [67]; preliminary studies suggest the material is a semiconductor with an indirect energy band gap.

5. Conclusions

For over a decade, indium chalcogenide nanomaterials continue to make significant contributions in the development of next-generation functional materials and devices, attributed to their unique properties which can be tuned easily using existing methods. As a result of their multiple crystallographic phases, in addition to the manipulation of the physical features such as morphology, indium chalcogenide nanomaterials remain of interest due to diversified opportunities which still need to be explored.

With the aid of computational modeling and related tools, it has become easier to identify application-specific objectives which guide the thought process when designing reaction protocols for nanomaterial fabrication. The current research-driven focus is on providing easy and efficient solutions to challenges associated with purity, quality and yield which affect the performance of the nanomaterial in desired applications. Hence, the recent literature reports provided in this book chapter have rather revisited classical methods of synthesis which are reputably known for producing high quality precursors, even though having received a lot of criticism over the years due to harsh and/or sensitive reaction protocols best executed by skilled personnel. Therefore, there is now and urgent need for the alternative routes such as the use of low-temperature decomposing single-source molecular precursors, which have been developed over the years, to be improved and incorporated in the fabrication of functional nanodevices.

In many literature reports, there continues to be an exacerbated use of 'non-toxic alternatives' and related terms whenever nanomaterials which do not contain heavy metals are presented. Novel and/or improved properties resulting from the physical changes of the material is a good indication that the nanomaterial could exhibit features and behavior different to the bulk counterpart, toxicity could be an example. Thus, an increasing trend on the interest of toxicity studies for nanomaterials is envisaged in the coming years.

Acknowledgements

The authors wish to thank the DST-NRF (South Africa) and RS-DFID (United Kingdom) for financial support.

Conflict of interest

We declare no conflict of interest.

Author details

Siphamandla C. Masikane and Neerish Revaprasadu*
University of Zululand, KwaDlangezwa, South Africa

*Address all correspondence to: revaprasadun@unizulu.ac.za

References

[1] Salvarli MS, Salvarli H. For Sustainable Development: Future Trends in Renewable Energy and Enabling Technologies. In: Al Qubeissi M, El-kharouf A, Soyhan HS, editors. Renewable energy - resources, challenges and applications. 1st ed. IntechOpen, 2020. p. 1-15. DOI: 10.5772/intechopen.91842

[2] Miller WF: Present and future nuclear reactor designs: Weighing the advantages and disadvantages of nuclear power with an eye on improving safety and meeting future needs. Journal of Chemical Education. 1993;70:109-114. DOI: 10.1021/ed070p109

[3] Mir M, Ishtiaq S, Rabia S, Khatoon M, Zeb A, Khan GM, Ur Rehman A, Ud Din F. Nanotechnology: From in vivo imaging system to controlled drug delivery. Nanoscale Research Letters. 2017;12:500. DOI: 10.1186/s11671-017-2249-8.

[4] Seegebarth B, Backhaus C, and Woisetschläger DM. The role of emotions in shaping purchase intentions for innovations using emerging technologies: A scenario-based investigation in the context of nanotechnology. Psychology & Marketing. 2019;36:844-862. DOI: 10.1002/mar.21228.

[5] Ko YH, Prabhakaran P, Choi S, Kim GJ, Lee C, Lee KS. Environmentally friendly quantum-dot color filters for ultra-high-definition liquid crystal displays. Scientific Reports. 2020;10:1-8. DOI: 10.1038/s41598-020-72468-8.

[6] Jin B, Zhai T. 2D Cadmium chalcogenides for optoelectronics. Chemical Research in Chinese Universities., 2020;36:493-503. DOI: 10.1007/s40242-020-0221-8.

[7] Lei H, Chen J, Tan Z, Fang G. Review of recent progress in antimony chalcogenide-based solar cells: Materials and devices. Solar RRL. 2019;3:1900026. DOI: 10.1002/solr.201900026.

[8] Rehman SU, Butt FK, Tariq Z, Li C. Tin-based novel cubic chalcogenides: A new paradigm for photovoltaic research. In: Kurinec SK, editor. Emerging photovoltaic materials: Silicon & beyond. Scrivener: Wiley; 2018. p. 141-163. DOI: 10.1002/9781119407690.ch5.

[9] Sharma Y, Srivastava P. Electronic, optical and transport properties of α-, β- and γ-phases of spinel indium sulphide: An ab initio study. Materials Chemistry and Physics. 2012;135:385-394. DOI: 10.1016/j.matchemphys.2012.04.064.

[10] Zavrazhnov AY, Naumov AV, Anorov PV, Goncharov EG, Sidei VI, Pervov VS. Tx phase diagram of the In-S system. Inorganic materials. 2006;42:1294-1298. DOI: 10.1134/S0020168506120028.

[11] Rajput N. Methods of preparation of nanoparticles-a review. International Journal of Advances in Engineering & Technology. 2015;7:1806-1811. ISSN: 22311963.

[12] Malik MA, Afzaal M, O'Brien P. Precursor chemistry for main group elements in semiconducting materials. Chemical reviews. 2010;110:4417-4446. DOI: 10.1021/cr900406f.

[13] Masikane SC, McNaughter PD, Lewis DJ, Vitorica-Yrezabal I, Doyle BP, Carleschi E, O'Brien P, Revaprasadu N. Important phase control of indium sulfide nanomaterials by choice of indium (III) xanthate precursor and thermolysis temperature. European Journal of Inorganic Chemistry. 2019:1421-1432. DOI: 10.1002/ejic.201900007.

[14] Afzaal M, Crouch D, O'Brien P. Metal-organic chemical vapor

deposition of indium selenide films using a single-source precursor. Materials Science and Engineering: B. 2005;116:391-394. DOI: 10.1016/j.mseb.2004.05.044.

[15] Souissi R, Bouguila N, Bendahan M, Fiorido T, Aguir K, Kraini M, Vázquez-Vázquez C, Labidi A. Highly sensitive nitrogen dioxide gas sensors based on sprayed β-In_2S_3 film. Sensors and Actuators B: Chemical. 2020;16:128280. DOI: 10.1016/j.snb.2020.128280.

[16] Kumtepe A, Altaf CT, Sahsuvar NS, Abdullayeva N, Koseoglu E, Sankir M, Sankir ND. Indium sulfide based photoelectrodes for all-vanadium photoelectrochemical redox flow batteries. ACS Applied Energy Materials. 2020;3:3127-3133. DOI: 10.1021/acsaem.9b02034.

[17] Gotoh T. Effect of heat treatments on the electronic properties of indium sulfide films. The European Physical Journal Applied Physics. 2020;89:20301. DOI: 10.1051/epjap/2020190240.

[18] Bchiri Y, Bouguila N, Kraini M, Souissi R, Vázquez-Vázquez C, López-Quintela MA, Alaya S. Investigation of the effect of S/In molar ratio on physical properties of sprayed In_2S_3 thin films. RSC Advances. 2020;10:21180-21190. DOI: 10.1039/D0RA02945A.

[19] Hariskos D, Hempel W, Menner R, Witte W. Influence of substrate temperature during In_xS_y sputtering on Cu(In, Ga)Se_2 buffer interface properties and solar cell performance. Applied Sciences. 2020;10:1052. DOI: 10.3390/app10031052.

[20] Tiss B, Bouguila N, Kraini M, Khirouni K, Vázquez–Vázquez C, Cunha L, Moura C, Alaya S. Electrical transport of sprayed In_2S_3: Ag thin films. Materials Science in Semiconductor Processing. 2020;114:105080. DOI: 10.1016/j.mssp.2020.105080.

[21] Rasool S, Saritha K, Reddy KR, Tivanov MS, Gremenok VF, Zimin SP, Pipkova AS, Mazaletskiy LA, Amirov II. Annealing and plasma treatment effect on structural, morphological and topographical properties of evaporated β-In_2S_3 films. Materials Research Express. 2020;7:016431. DOI: 10.1088/2053-1591/ab6a5b.

[22] Tu CL, Lin KI, Pu J, Chung TF, Hsiao CN, Huang AC, Yang JR, Takenobu T, Chen CH. CVD growth of large-area InS atomic layers and device applications. Nanoscale. 2020;12:9366-9374. DOI: 10.1039/d0nr01104e.

[23] Luo D, Zhou B, Guo B, Gao P, Zheng L, Zhang X, Cui S, Zhou H, Zhou Y, Liu Y. Solution-processable two-dimensional ultrathin nanosheets induced by self-assembling geometrically-matched alkane. Nano Energy. 2020;72:104689. DOI: 10.1016/j.nanoen.2020.104689.

[24] Park JH, Chung TM, Park BK, Kim CG. Indium complexes with aminothiolate ligands as single precursors for indium chalcogenides. Inorganica Chimica Acta. 2020;505:119504. DOI: 10.1016/j.ica.2020.119504

[25] Wang L, Guan Z, Tang A. Multinary copper-based chalcogenide semiconductor nanocrystals: Synthesis and applications in light-emitting diodes and bioimaging. Journal of Nanoparticle Research. 2020;22:1-20. DOI: 10.1007/s11051-019-4724-x.

[26] Marin R, Skripka A, Huang YC, Loh TA, Mazeika V, Karabanovas V, Chua DH, Dong CL, Canton P, Vetrone F. Influence of halide ions on the structure and properties of copper indium sulphide quantum dots. Chemical Communications. 2020;56:3341-3344. DOI: 10.1039/C9CC08291C

[27] Pawar AS, Masikane SC, Mlowe S, Garje SS, Revaprasadu N.

Preparation of CdS nanoparticles from thiosemicarbazone complexes: Morphological influence of chlorido and iodido ligands. European Journal of Inorganic Chemistry. 2016;2016:366-372. DOI: 10.1002/ejic.201501125.

[28] Masikane SC, Mlowe S, Pawar AS, Garje SS, Revaprasadu N. Cadmium chloride and cadmium iodide thiosemicarbazone complexes as single source precursors for CdS nanoparticles. Russian Journal of Inorganic Chemistry. 2019;64:1063-1071. DOI: 10.1134/S0036023619080072.

[29] Masikane SC, Mlowe S, Gervas C, Revaprasadu N, Pawar AS, Garje SS. Lead (II) halide cinnamaldehyde thiosemicarbazone complexes as single source precursors for oleylamine-capped lead sulfide nanoparticles. Journal of Materials Science: Materials in Electronics. 2018;29:1479-1488. DOI: 10.1007/s10854-017-8056-2.

[30] Fuhr A, Yun HJ, Crooker SA, Klimov VI. Spectroscopic and magneto-optical signatures of Cu^{1+} and Cu^{2+} defects in copper indium sulfide quantum dots. ACS nano. 2020;14:2212-2223. DOI: 10.1021/acsnano.9b09181.

[31] Kays JC, Saeboe AM, Toufanian R, Kurant DE, Dennis AM. Shell-free copper indium sulfide quantum dots induce toxicity *in vitro* and *in vivo*. Nano Letters. 2020;20:1980-1991. DOI: 10.1021/acs.nanolett.9b05259.

[32] Lian ZJ, Lin TY, Yao CX, Su YL, Liao SH, Wu SM. Staphylococcus aureus strains exposed to copper indium sulfide quantum dots exhibit increased tolerance to penicillin G, tetracycline and ciprofloxacin. New Journal of Chemistry. 2020;44:6533-6542. DOI: 10.1039/c9nj05748j.

[33] Jiao M, Li Y, Jia Y, Li C, Bian H, Gao L, Cai P, Luo X. Strongly emitting and long-lived silver indium sulfide quantum dots for bioimaging: Insight into co-ligand effect on enhanced photoluminescence. Journal of Colloid and Interface Science. 2020;565:35-42. DOI: 10.1016/j.jcis.2020.01.006.

[34] Naik SD, Apte SK, Garaje SN, Sethi YA, Shinde MD, Arbuj SS, Kale BB, Sonawane R. Facile template free approach for the large scale solid phase synthesis of nanocrystalline XIn_2S_4 (X= Cd/Zn) and it's photocatalytic performance for H_2 evolution. New Journal of Chemistry. 2020;44:9634-9646. DOI: 10.1039/d0nj01323d.

[35] Kale BB, Baeg JO, Lee SM, Chang H, Moon SJ, Lee CW. $CdIn_2S_4$ nanotubes and "Marigold" nanostructures: a visible-light photocatalyst. Advanced Functional Materials. 2006;16:1349-1354. DOI: 0.1002/adfm.200500525.

[36] Chaudhari NS, Bhirud AP, Sonawane RS, Nikam LK, Warule SS, Rane VH, Kale BB. Ecofriendly hydrogen production from abundant hydrogen sulfide using solar light-driven hierarchical nanostructured $ZnIn_2S_4$ photocatalyst. Green Chemistry. 2011;13:2500-2506. 10.1039/C1GC15515F.

[37] Lisensky G, McFarland-Porter R, Paquin W, Liu K. Synthesis and Analysis of Zinc Copper Indium Sulfide Quantum Dot Nanoparticles. Journal of Chemical Education. 2020;97:806-812. DOI: 10.1021/acs.jchemed.9b00642.

[38] Preeyanka N, Dey H, Seth S, Rahaman A, Sarkar M. Highly efficient energy transfer from a water soluble zinc silver indium sulphide quantum dot to organic J-aggregates. Physical Chemistry Chemical Physics. 2020;22:12772-12784. DOI: 10.1039/d0cp01845g.

[39] Han G, Chen ZG, Drennan J, Zou J. Indium selenides: structural characteristics, synthesis and their

thermoelectric performances. Small. 2014;10:2747-2765. DOI: 10.1002/smll.201400104.

[40] Boukhvalov DW, Gürbulak B, Duman S, Wang L, Politano A, Caputi LS, Chiarello G, Cupolillo A. The advent of indium selenide: Synthesis, electronic properties, ambient stability and applications. Nanomaterials. 2017;7:372. DOI: 10.3390/nano7110372.

[41] Mamedov RM. A new way of obtaining ultrathin films of indium selenide. Russian Journal of Physical Chemistry A. 2020;94:1272-1275. DOI: 10.1134/S0036024420060163.

[42] Hao S, Yan S, Wang Y, Xu T, Zhang H, Cong X, Li L, Liu X, Cao T, Gao A, Zhang L. Edge-epitaxial growth of InSe nanowires toward high-performance photodetectors. Small. 2020;16:1905902. DOI: 10.1002/smll.201905902.

[43] Zhang X, Lee S, Bansal A, Zhang F, Terrones M, Jackson TN, Redwing JM. Epitaxial growth of few-layer β-In_2Se_3 thin films by metalorganic chemical vapor deposition. Journal of Crystal Growth. 2020;533:125471. DOI: 10.1016/j.jcrysgro.2019.125471.

[44] Shi H, Li M, Shaygan Nia A, Wang M, Park S, Zhang Z, Lohe MR, Yang S, Feng X. Ultrafast electrochemical synthesis of defect-free In_2Se_3 flakes for large-area optoelectronics. Advanced Materials. 2020;32:1907244. DOI: 10.1002/adma.201907244.

[45] Hai T, Xie G, Qiao Z, Qin Z, Ma J, Sun Y, Wang F, Yuan P, Ma J, Qian L. Indium selenide film: a promising saturable absorber in 3- to 4-μm band for mid-infrared pulsed laser. Nanophotonics. 2020;9:2045-2052. DOI: 10.1515/nanoph-2020-0068.

[46] Andres-Penares D, Canet-Albiach R, Noguera-Gomez J, Martínez-Pastor JP, Abargues R, Sánchez-Royo JF. Two-dimensional indium selenide for sulphur vapour sensing applications. Nanomaterials. 2020;10:1396. DOI: 10.3390/nano10071396.

[47] Chen F, Cui A, Wang X, Gao C, Xu L, Jiang K, Zhang J, Hu Z, Chu J. Lattice vibration characteristics in layered InSe films and the electronic behavior of field-effect transistors. Nanotechnology. 2020;31:335702. DOI: 10.1088/1361-6528/ab8df1.

[48] Sangwan VK, Kang J, Hersam MC. Thickness-dependent charge transport in exfoliated indium selenide vertical field-effect transistors. Applied Physics Letters. 2019;115:243104. DOI: 10.1063/1.5128808.

[49] Chen YH, Cheng CY, Chen SY, Rodriguez JS, Liao HT, Watanabe K, Taniguchi T, Chen CW, Sankar R, Chou FC, Chiu HC. Oxidized-monolayer tunneling barrier for strong Fermi-level depinning in layered InSe transistors. npj 2D Materials and Applications. 2019;3:1-7. DOI: 10.1038/s41699-019-0133-3.

[50] Wang F, Jiang J, Liu Q, Zhang Y, Wang J, Wang S, Han L, Liu H, Sang Y. Piezopotential gated two-dimensional InSe field-effect transistor for designing a pressure sensor based on piezotronic effect. Nano Energy. 2020;70:104457. DOI: 10.1016/j.nanoen.2020.104457.

[51] Liao Y, Shan Y, Wu L, Xiang Y, Dai X. Liquid-exfoliated few-Layer InSe nanosheets for broadband nonlinear all-optical applications. Advanced Optical Materials. 2020;8:1901862. DOI: 10.1002/adom.201901862.

[52] Long H, Liu S, Wen Q, Yuan H, Tang CY, Qu J, Ma S, Qarony W, Zeng LH, Tsang YH. In_2Se_3 nanosheets with broadband saturable absorption used for near-infrared femtosecond laser mode locking.

Nanotechnology. 2019;30:465704. DOI: 10.1088/1361-6528/ab33d2.

[53] Yang D, Zhu Q, Sun X, Chen C, Guo W, Yang G, Han B. Electrosynthesis of a defective indium selenide with 3D structure on a substrate for tunable CO_2 electroreduction to syngas. Angewandte Chemie. 2020;132:2374-2379. DOI: 10.1002/ange.201914831.

[54] Mech RK, Mohta N, Chatterjee A, Selvaraja SK, Muralidharan R, Nath DN. High responsivity and photovoltaic effect based on vertical transport in multilayer α-In_2Se_3. physica status solidi (a). 2020;217:1900932. DOI: 10.1002/pssa.201900932.

[55] Hossain J, Julkarnain M, Mondal BK, Newaz MA, Khan KA. Unveiling the Electrical and thermoelectric properties of highly degenerate indium selenide thin films: Indication of In_3Se_4 phase. Materials Research Express. 2019;6:126421. DOI: 10.1088/2053-1591/ab5ac1.

[56] Sousa FL, Souza BA, Jesus AC, Azevedo WM, Mansur HS, Freitas DV, Navarro M. Aqueous electrosynthesis of silver indium selenide nanocrystals and their photothermal properties. Green Chemistry. 2020;22:1239-1248. DOI: 10.1039/c9gc03647d.

[57] Yang J, Zhou Z, Fang J, Wen H, Lou Z, Shen G, Wei Z. Magnetic and transport properties of a ferromagnetic layered semiconductor $MnIn_2Se_4$. Applied Physics Letters. 2019;115:222101. DOI: 10.1063/1.5126233.

[58] Hu X, Du L, Wang Y, Lahtinen J, Yao L, Ren Z, Sun Z. Raman fingerprints and exciton-phonon coupling in 2D ternary layered semiconductor InSeBr. Applied Physics Letters. 2020;116:163105. DOI: 10.1063/1.5143119.

[59] Regmi G, Ashok A, Chawla P, Semalti P, Velumani S, Sharma SN, Castaneda H. Perspectives of chalcopyrite-based CIGSe thin-film solar cell: a review. Journal of Materials Science-Materials in Electronics. 2020;31:7286-7314. DOI: 10.1007/s10854-020-03338-2

[60] Marvan P, Mazánek V, Sofer Z. Shear-force exfoliation of indium and gallium chalcogenides for selective gas sensing applications. Nanoscale. 2019;11:4310-7. DOI: 10.1039/c8nr09294j.

[61] Fu Y, Zhou J, Zou HH, Almeida Paz FA, Liu X, Fu L. Unique two-dimensional indium telluride templated by a rare wheel-shaped heterobimetallic Mn/In cluster. Inorganic Chemistry. 2020;59:5818-2582. DOI: 10.1021/acs.inorgchem.0c00526.

[62] Hsin CL, Huang CW, Wu MH, Cheng SY, Pan RC. Synthesis and thermoelectric properties of indium telluride nanowires. Materials Research Bulletin. 2019;112:61-65. DOI: 10.1016/j.materresbull.2018.12.006.

[63] Vallem S, Bangera KV, Shivakumar GK. Enhanced thermoelectric power of Al and Sb doped In_2Te_3 thin films. Materials Science in Semiconductor Processing. 2019;93:366-370. DOI: 10.1016/j.mssp.2019.01.025.

[64] Zhu H, Zhang B, Wang G, Peng K, Yan Y, Zhang Q, Han X, Wang G, Lu X, Zhou X. Promoted high temperature carrier mobility and thermoelectric performance of InTe enabled by altering scattering mechanism. Journal of Materials Chemistry A. 2019;7:11690-11698. DOI: 10.1039/C9TA00475K.

[65] Ahmed F, Tsujii N, Matsushita Y, Sauerschnig P, Mori T. Influence of slight substitution (Mn/In) on thermoelectric and magnetic properties in chalcopyrite-type $CuInTe_2$. Journal of Electronic Materials. 2019;48:4524-4532. DOI: 10.1007/s11664-019-07234-2.

[66] Zhong Y, Luo Y, Li X, Cui J. Silver vacancy concentration engineering leading to the ultralow lattice thermal conductivity and improved thermoelectric performance of $Ag_{1-x}InTe_2$. Scientific reports. 2019;9:1-8. DOI: 10.1038/s41598-019-55458-3.

[67] Bouchenafa M, Benmakhlouf A, Sidoumou M, Bouhemadou A, Maabed S, Halit M, Bentabet A, Bin-Omran S, Khenata R, Al-Douri Y. Theoretical investigation of the structural, elastic, electronic, and optical properties of the ternary tetragonal tellurides $KBTe_2$ (B= Al, In). Materials Science in Semiconductor Processing. 2020;114:105085. DOI: 10.1016/j.mssp.2020.105085.

Chapter 3

Features of the Ionic State of Indium in Perchlorate Solutions and the Physicochemical Properties of Indium Perchlorate

Boris Radionov Konstantinovich and Ilya Svirsky Anatolievich

Abstract

The possibility of the formation of India perchlorate and mixed ligand complexes has been studied. It is assumed that during the extraction of indium in aqueous solutions of hydrochloric acid at various concentrations and constant ionic strength maintained by the addition of perchloric acid, indium is present in the organic phase in the form of ionic aggregates $HClO_4 \cdot HInCl_4$. To study the hydration of Ions, the method of electrical conductivity was carried out in the "$In(ClO_4)_3$ - $HClO_4$ - H_2O" system. The structure of aqueous solutions of India perchlorates was determined by IR spectroscopy. The structures of trivalent perchlorate India have been established.

Keywords: perchlorate ion, indium, gallium, complexation, ionic state, NMR, IR, extraction, coordination number

1. Ionic state of indium in perchlorate solutions

It is known that perchlorate ion does not form complex compounds even with extremely strong complexing metal ions [1]. The difficulty in preparing perchlorate metal complexes is the lack of suitable solvents. Water molecules and most non-aqueous donor media displace such a weak ligand as the ClO_4^- ion from the inner sphere of the compound. The formation of a coordination bond between the perchlorate ion and the complexing cation is possible only in acceptor or very weakly donating solvents. The successful synthesis of perchloro-complexes in work [2] is associated with the use of anhydrous perchloric acid as a reaction medium.

These fairly well known provisions are fully confirmed for indium perchlorate, which is proved by various research methods [3–19]. The absence of coordination interaction of acid ions in the "In^{3+} - ClO_4^- – H_2O" system is indicated by the spectra of Raman scattering [5–11]. IR spectra [12] including near infrared [13–15] and spectroscopy of disturbed total internal reflection [13, 16], NMR signals [17, 18]. However, partial formation of ion pairs between them is allowed [8, 18].

In the Raman spectra of the perchlorate solution, only the lines of the ClO_4^- anion and the aquocomplexes of the In^{3+} ion ·aq were found [6]. Raman and IR spectroscopic studies of indium hydration in perchlorate solutions revealed octahedral hexaaquocation $[In(H_2O)_6]^{3+}$. According to the data of the Raman spectra, the aquocomplex is stable in acidic perchlorate solutions, and in the studied concentration range, neither inner sphere nor hydroxostructures are formed [9].

There is no evidence of the complexing effect of the highly concentrated electrolyte $NaClO_4$ 1.1÷9.1 M and by electrophoresis [19]. Spectrophotometrically in the ultraviolet region of the spectrum, there were no significant signs of direct interaction in aqueous solutions of $In(ClO_4)_3$ salts [20]. Also, no intrinsic complexation influencing the activity of In^{3+} was reliably detected up to a sodium perchlorate concentration of 16 mol/kg [19, 21] and with an $In(ClO_4)_3$ content of more than 4 M [6, 9].

Similar results, denying the likelihood of coordination interaction of In^{3+} with ClO_4^-, were obtained when studying the solvation of cations by NMR on H^1 and P^{31} nuclei in aqueous-organic mixtures [22].

The proton-magnetic resonance measurement of the coordination number of perchlorate water-organic systems testifies to the six fold coordination of water molecules around the indium cation and to the absence of strong evidence of contact ion pairing [23, 24]. It was not possible to detect the binding of indium to perchlorate ions ClO_4^- by special experiments carried out in mixed water-non-aqueous mixtures and organic media [25].

The absence of complex compounds of indium with perchlorate ion is also indicated by the data of liquid extraction, in particular, three n-octylamine does not extract indium from a solution with an $HClO_4$ concentration less than 2 M [4]. At the same time, from concentrated solutions of perchloric acid for cationic reagents - alkylphosphoric acids (for example, di-2-ethylhexylphosphoric acid; **Figure 1**) - a possible extractable form of indium compounds is the complex cation $[InClO4]^{2+}$ [26]. It is also assumed that during the extraction of indium in aqueous solutions of hydrochloric acid at its various concentrations and constant ionic strength maintained by the addition of perchloric acid, indium is present in the organic phase in the form of ionic aggregates $HClO_4{\cdot}HInCl_4$ [27]. The latter formation is possibly caused by the coextractability of perchloric acid due to its higher extraction affinity as compared to hydrochloric acid. In the aqueous phase, there are only chloride complexes, which are subjected to extraction with various oxygen-containing solvents.

The likelihood of the formation of perchlorate and mixed-ligand complexes also applies to extraction systems based on 4-methyl-2-pentanone and 4-methyl-2-pentanol, the extractable forms of which are compounds of the composition $[In(ClO_4)_3]$, $[InHal(ClO_4)_2]$, $[InHal_2(ClO_4)]$, where bromine, iodide, or thiocyanate ion is present

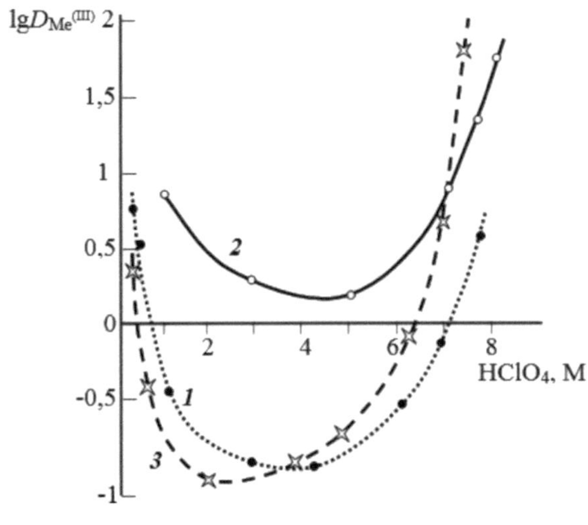

Figure 1.
Extraction of III subgroup metal ions with di (2-ethylhexyl) phosphoric acid depending on the concentration of perchloric acid: 1 - Ga (III), 2 - In (III), 3 - Tl (III) [26].

as the halide and pseudohalide of the ligand. The aqueous phase is characterized by a constant ionic strength of 4.0 due to the salt background of sodium perchlorate and salts of NaBr, NaJ or NaSCN of variable concentrations. The role of $NaClO_4$ in liquid distribution is reduced not only to a change in the activity and composition of the solution, but also to the extraction of ion pairs of charged indium complexes, which greatly increases in the presence of perchlorate ions [28].

An example of the special effect of perchlorate ion on the extraction behavior of indium (III) is its interaction in a mixture of extractants 1-phenyl-3-methyl-4-acylpyrazol-5 (PhMAPr) with tri-n-phosphine oxide (TOPhO) in toluene [29] and PhMAPr with a base salt — Aliquat-336 [29]. If the reaction of interfluid distribution in a perchlorate medium proceeds according to the equation (1):

$$In^{3+} + 3HPhMAPr_{(o)} \leftrightarrow In\,(PhMAPr)_{3(o)} + 3H^{+} \text{ with high } logK_{ex} = 1.50, \quad (1)$$

and from aqueous solutions of H(Na)Cl, ClO_4 during extraction with a mixture PhMAPr and TOPhO have a synergistic effect and the process is described by the equation (2):

$$[InCln]^{3-n} + (3-n)HPhMAPr_o + 2TOPhO_o \leftrightarrow InCl_n\,(PhMAPr)_{3-n} \cdot 2TOPhO_o + (3-n)H^{+} \quad (2)$$

where n = 0–3, then in the case of perchlorate salt Aliquat-336, this phenomenon is not observed due to its finding in the form of $R_4N^{+}ClO_4^{-}$, although when using in the same system high-molecular-weight ammonium in chloride and nitrate forms, the effect of synergism is not only preserved, but also increased.

When studying the coordination properties of polyvalent metal ions, including indium, by titrimetric and ion-exchange methods, only weak signs of complexation in perchloric acid were noted [30]

The cation-exchange behavior of indium (III) ions in the presence of perchloric acid and an organic solvent in the form of methyl alcohol and acetone also indicates the absence of fundamental changes in the ionic state of indium in such a sorption system. Both in the presence and in the absence of these solvents, the indium ion interfacial distribution indices in the perchloric acid medium are higher than in the hydrochloric acid medium (picture 2) [31]. Thus, if the coordination properties of such acid ligands as halides, sulfate, thiocyanate [32], and to some extent nitrite [33] in aqueous-organic mixtures can be activated, which leads to the formation of more acid-saturated complexes, then the perchlorate ion under these conditions in relation to indium, as a complexing agent, remains very passive (**Figure 2**).

From other alternative points of view, attention is drawn to the study of the effect of the salting-out electrolyte on the distribution coefficient of indium during the extraction of its tenoyl-three-fluoroacetonate complex with benzene from aqueous perchlorate solutions, where the presence of complexation of indium with the perchlorate ion was confirmed.

The revealed relationship between the salting-out parameter and the stability of indium complexes with various anions formed in the aqueous phase indicates that the strength of the acidic complexes by the nature of the ligand changes in the series of salts: NaCl > > $NaNO_3$ > $NaClO_4$ [34].

A characteristic feature of indium perchlorate solutions, like other indium salts, is the extremely high viscosity of aqueous solutions of $In(ClO_4)_3$, due to the increased degree of hydration of the three-charged indium ion [6]. A simple substance with the composition $In(ClO_4)_3 \sim 8H_2O$ crystallizes from concentrated aqueous solutions of indium perchlorate [11], and in dilute perchloric acid, indium is in the form of the cation $[In(H_2O)_6]^{3+}$.

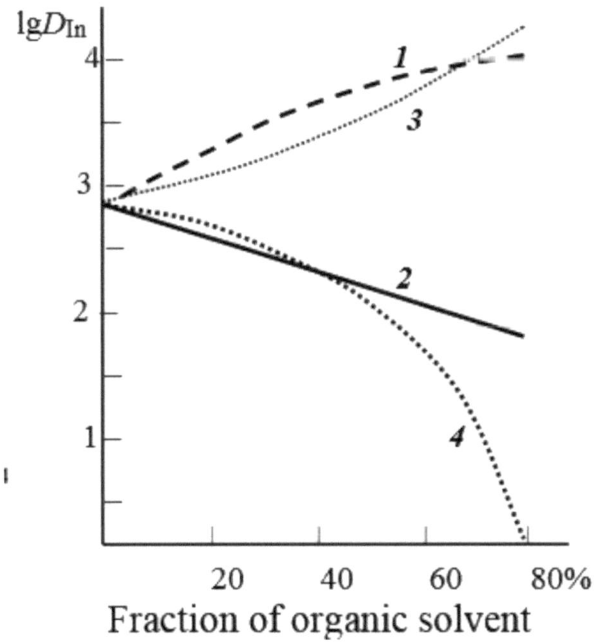

Figure 2.
Dependence of the distribution coefficients of indium in 0.1 M $HClO_4$ (1.3) and 0.1 M HCl (2, 4) on the content of the organic solvent: 1 and 2 - methanol; 3 and 4 - acetone [31].

2. Study of ion hydration by the method of electrical conductivity in the system

"$In(ClO_4)_3$ - $HClO_4$– H_2O" shows that the total number of water molecules oriented by the cation is n_{In3+} = ~ 9, which includes water molecules that make up the near and far environment of the cation. The decrease in n_{In3+} relative to the maximum value equal to 12 reflects the approach of the hydration and anionic environment of indium to what is characteristic of crystalline hydrates, and indium perchlorate crystallizes from an aqueous solution with eight water molecules. In the series of cations III of the Al^{3+}-In^{3+} subgroup, there is a tendency to a decrease in n_{Mes+} with an increase in the crystal chemical radius. The reason for this is the decrease in the polarizing ability of the cations. Consequently, this leads to a weakening of the bonds of near and outer-sphere water molecules with a probable increase in the residence time of anions near the $[Me(H_2O)_6]^{3+}$ cation [35].

The solubility of indium perchlorate salt at 0° and 25°C is 3.23 and 3.53 mol/kg H_2O, respectively. More accurately speaking about the macroscopic properties, about the possible composition of crystalline hydrates precipitating into the bottom phase, about the processes preceding crystal formation, allow the polytherms of solubility [36]. Their analysis shows that there is a qualitative similarity in the solubility polytherms of the aluminum subgroup perchlorates. The value of the molar fraction of water unbound to salt in the eutectic for perchlorates is constant (N_w = 0.94 ± 0.1). The freezing temperatures of eutectic compositions are close to −28 ÷ −31°C and all curves have an inflection - a maximum that corresponds to the existence of a chemical compound. In the case of indium perchlorate, this is a 28-aqueous crystalline hydrate. In addition, after reaching the concentration of 2.8 mol/kg H_2O, the phenomenon of glass transition of this solution is observed. However, the position of the eutectic concentration on the solubility polytherm depends not so much on the nature of the cation as on the number of water molecules that are part of the crystalline hydrate. The reason for the existence of compounds enriched with water in the eutectic should be sought in the fact that the stable bridges formed as a result of local hydrolysis (cation ··· OH^- ··· H_3O^+) are able to retain a large amount of water. Thus, providing the possibility of the formation of high-water crystalline hydrates at temperatures below 0°C. However, as the solution concentrates, structurally forced processes begin. Rearrangement of one dominant structure into another containing a smaller number of water molecules in the cybotactic group. These processes are accompanied by glass transition (**Figure 3**).

The structure of aqueous solutions of indium perchlorates and the state of water were studied by IR spectroscopy in the region of the first overtone of water (NIR spectra) using chemometric analysis to the concentration dependences of the spectra. It is noted that hydrolysis occurs in indium (III) perchlorate solutions, and due to the high enthalpy of hydration of In3 +, exclusively solvate-separated ion pairs are formed. In other words, the formation of complexes of the composition In^{3+} (H_2O) n $(ClO_4)^-$ occurs, where n = 1–2 with a decrease in the symmetry of the ClO_4^- ion: Td → C_{3v} → C_{2v} with an increase in the concentration of indium in solutions acidified with perchloric acid [14, 15].

Perchlorate ion has the shape of a regular tetrahedron. Its hydration shell consists of eight water molecules. The distribution of these shells around the ClO_4^- ion is uniform. Presumably, it is low hydrating capacity [37].

Indium perchlorate salt is a hygroscopic substance readily soluble in water and it can be obtained by dissolving indium metal in perchloric acid at a moderate temperature. Its anhydrous modification crystallizes because of the interaction of silver perchlorate and indium chloride in a methanol solution. There are also several forms of indium hydroxopochlorates, which are precipitated in their solutions in the presence of sodium perchlorate [38]. At pH 4.82, a salt of the composition $In_4OH(ClO_4)_{11}$ was isolated, at pH 3.77 - $In_2OH(ClO_4)_5$, and in the pH range 2.56–3.14 - $In_4(OH)_3(ClO_4)_9$.

Determination of the apparent ionic volumes of particles by measuring the density of indium (III) perchloric acid solutions showed [39] that, depending on the degree of dilution, the formation of a compound of the composition $[In(H_2O)_5ClO_4]^{2+}$. However, there are no convincing reasons to consider the possible entry of the perchlorate ion into the inner sphere of the indium aquion.

According to X-ray diffraction data, the structure of a hydrated indium ion (III) in a 3 M aqueous solution of perchlorate is an aquo-complex, in which the first hydration shell contains six water molecules with a bond length of 2.15 ± 0.03 Å between In^{3+} and nearby H_2O [40].

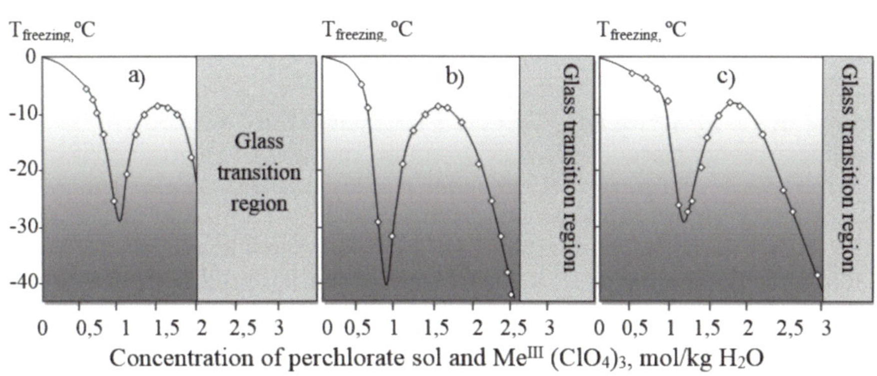

Figure 3.
Perchlorate solubility polytomes: (a) Al $(ClO_4)_3$, (b) Ga $(ClO_4)_3$, (c) In $(ClO_4)_3$ [36].

To establish the structure of trivalent indium perchlorate, the crystalline salt was obtained by dissolving indium (III) oxide in an equimolar amount of perchloric acid on heating– $[In(H_2O)_6](ClO_4)_3$. By isothermal evaporation at room temperature from an aqueous solution of In $In(ClO_4)_3$, crystals $[In(H_2O)_5](ClO_4)_3$ unstable in air were isolated. X-ray structural analysis of two forms of indium perchlorate - $[In(H_2O)_6](ClO_4)_3$ and single crystal $[In(H_2O)_6](ClO_4)_3{\cdot}3H_2O$ indicates that both these structures are formed by hexaaquocation $[In(H_2O)_6]^{3+}$ and perchlorate ions. In this case, $[In(H_2O)_6](ClO_4)_3{\cdot}3H_2O$ contains crystallization water molecules. But the structure of $[In(H_2O)_6](ClO_4)_3$ is cubic, and $[In(H_2O)_6](ClO_4)_3{\cdot}3H_2O$ is rhombic, each of them with the corresponding definite geometric parameters, in which the aquated indium complexes are close to the ideal octahedron. Special attention should be paid to the hydration state of indium perchlorates. In all likelihood, the hydration composition of its salt is not eight, but nine-water, consisting of six-coordinated water molecules and three crystallization ones. At the same time, the existence of a six-water modification unstable in air that does not contain molecules of crystallized water is possible. However, since the compound is highly hygroscopic, it is possible that partial dehydration occurs in the processes of pre-preparation of samples for chemical analysis $[In(H_2O)_6](ClO_4)_3{\cdot}3H_2O$ [41].

The available X-ray diffraction measurements carried out for perchlorate solutions of indium (III) at temperatures from 25 to 250° C under water vapor pressure indicate that the distance In^{3+} – oxygen (water) in the octahedral configuration of the aqua complex remains constant and equal to 2.14 ± 0.01 Å. In 1.0 M $HClO_4$ solutions, the formation of the ion pair In^{3+} – perchlorate $[InClO_4(H_2O)_5]^{2+}$ is observed, in which the distance In^{3+} – chlorine (perchlorate) changes from 3.14 to 3.12 Å with an increase in temperature to 250° C. At the same time, no noticeable destruction of the octahedral hydration shell of the aqua-ion of indium at temperatures up to 300° C was recorded [42].

In addition to the crystalline hydrate of the composition $In(ClO_4)_3{\cdot}8H_2O$, isolated from an aqueous solution, the corresponding dimethylforamide solvate $In(ClO_4)_3{\cdot}6(CH_3)_2NCHO$ was isolated by the preparative method of double recrystallization due to the solvation effect caused by the peculiarities of the chemical interaction of indium ions with molecules of an organic solvent [43].

The anhydrous salt of indium (III) perchlorate is obtained in the course of the exchange reaction between $InCl_3$ or InJ_3 with $AgClO_4$ in methanol solution [44].

Sequential thermal dehydration of indium perchlorate heptahydrate leads to the decomposition of the salt, the final product of which is indium trioxide [45].

When studying the effect of the composition and concentration of the water-salt system on the EMF of the cell $Hg_{(liquid)}/Hg_2Cl_{2(solid.)}/NaCl_{(saturated.)}/NaAni//$ $Me(ClO_4)_n$, $In(ClO_4)_3/In_{(solid.)}$, где Men^+ = Li, Na, Mg; Na*Ani* = 5416 M NaCl, 4,0 M $NaNO_3$, 3,72 M and 6,72 M $NaClO_4$ it was shown that the nature of the salts that form the salt bridge and their amount have a definite effect on the EMF of the cell. The greatest difference in the EMF values was noted when comparing the units of 4.0 M $NaNO_3$ and 6.72 M $NaClO_4$, and in the case of 3.72 M $NaClO_4$, the values are very close. The chemical composition and concentration of the salt background have the greatest effect on the emf and the activity coefficient of the In3 + cation (γ ±): the γ ± values rapidly increase (by 5–6 orders of magnitude) with a change in the magnesium salt content from 0.5 to 4 M and up to 13 M sodium salts. Depending on the nature of the salt background, γ ± is ranked in the order $Mg(ClO_4)_2 > LiClO_4 > NaClO_4$ [21].

IR spectroscopy of the synthesized series of compounds of the cationic type $[In(Lig)_n](ClO_4)_3$, when n = 6, 4, 3, and 2, and the ligands are urea, dimethyl sulfoxide, Ph_3P, Ph_3PO_4, 2,2′-*Dipy*, *En*, 2,2,2′-three-pyridyl and others, indicates that the perchlorate ion is a part of the anionic part of the complex structure [46]. By means of an exchange reaction in acetone solutions, hexa aqua salt $[In(H_2O)_6]$ $(ClO_4)_3$ under the action of dimethyl sulfoxide or dimethylformamide is easily converted into solvates $[In(DMSO)_6](ClO_4)_3$ and $[In(DMFA)_6](ClO_4)_3$ [47]. Subsequently, X-ray structural studies established the structure of the crystal complex of indium hexa (dimethylsulfoxide) perchlorate (III) [In(*ДМСО*)$_6$]$(ClO_4)_3$ [48].

Thus, there are clearly no compelling reasons to consider the existence of complexes of indium with perchlorate ion, especially its anionic forms, as possible, and as a consequence, the indium ion in such a medium is a simple hydrated cation. Apparently, under certain conditions, it is possible to admit the formation of ion pairs, where the perchlorate ligand is localized in the outer sphere with respect to the central atom, i.e. attached to all aquo-ion.

Therefore, in the practice of studying the complexing properties of indium, as well as most other metal ions, salt perchlorate and perchloric acid are traditionally used to create a given ionic strength of a solution and as a source of acid anions that do not form complexes. And also in those cases when it is necessary to assess the interaction of cations, for example, with water, not complicated by the formation of their acidic complexes.

The macro-properties of aqueous solutions of electrolytes of the gallium subgroup are significantly influenced by the peculiarities of the interaction of metal cations with water. The regularities of changes in such characteristics as the bulk properties of perchlorate solutions, isobaric expandability, adiabatic compressibility and heat capacity (isobaric and isochoristic) [49, 50] were analyzed from the point of view of the specificity of reactions of metal cations with dipoles of water molecules, depending on their electronic structure p- and d-metals of the third group [51].

It is quite natural that the studied macroproperties change nonmonotonically in the gallium subgroup. In this case, the type of non-monotony corresponds to the unequal sensitivity of each property to one or another result of interaction with water. It can be seen from **Figure 4** that when passing from the dependences $\Delta F_к$ to the dependences ΔV, ΔF_E and ΔC_p, the direction of the break in the Ga – In – Tl line gradually changes to the opposite. It follows from this that the ratio of the factors of water binding and its polarization is different for different properties. The factor of water binding predominates in the influence of the $\Delta F_к$ compressibility, which leads to the appearance of a rather sharp maximum on the "$\Delta F_к$ - $N_{Ga\text{-}In\text{-}Tl}$" contour at the point of the least vigorously hydrated indium ion. In the case of the characteristics of the specific volume ΔV, the effect of compression of the bound water is partly compensated by the formation of openwork outer-sphere associates, as a result of which the maximum on the ΔV curve disappears.

The course of the dependence "ΔF_E - $N_{Ga\text{-}In\text{-}Tl}$" is antibate to the curve "$\Delta F_к$ - $N_{Ga\text{-}In\text{-}Tl}$", which may, firstly, be a consequence of the unequal degree of compaction of water molecules during hydration, and the greater the degree of compaction, the more ΔF_E, and, secondly, is due to different degrees of water polarization and localized hydrolysis of cations. Finally, the dependence "ΔC_p - $N_{Ga\text{-}In\text{-}Tl}$" demonstrates the absolute predominance of the effects of water polarization and localized hydrolysis over the influence of differences in the degree of water binding by cations, since the last of these factors should have caused the reverse course of the curve. Such a significant positive effect of water polarization and localized hydrolysis on heat capacity is apparently caused by corresponding changes in the set of frequencies of intra- and intermolecular vibrations of the aqua complex as a result of polarization, as well as redistribution of "O – H" bonds and hydrogen bonds due to local hydrolysis.

The comparisons made refer to the region of relatively dilute solutions. With an increase in concentration, the pattern of changes in properties from subgroup

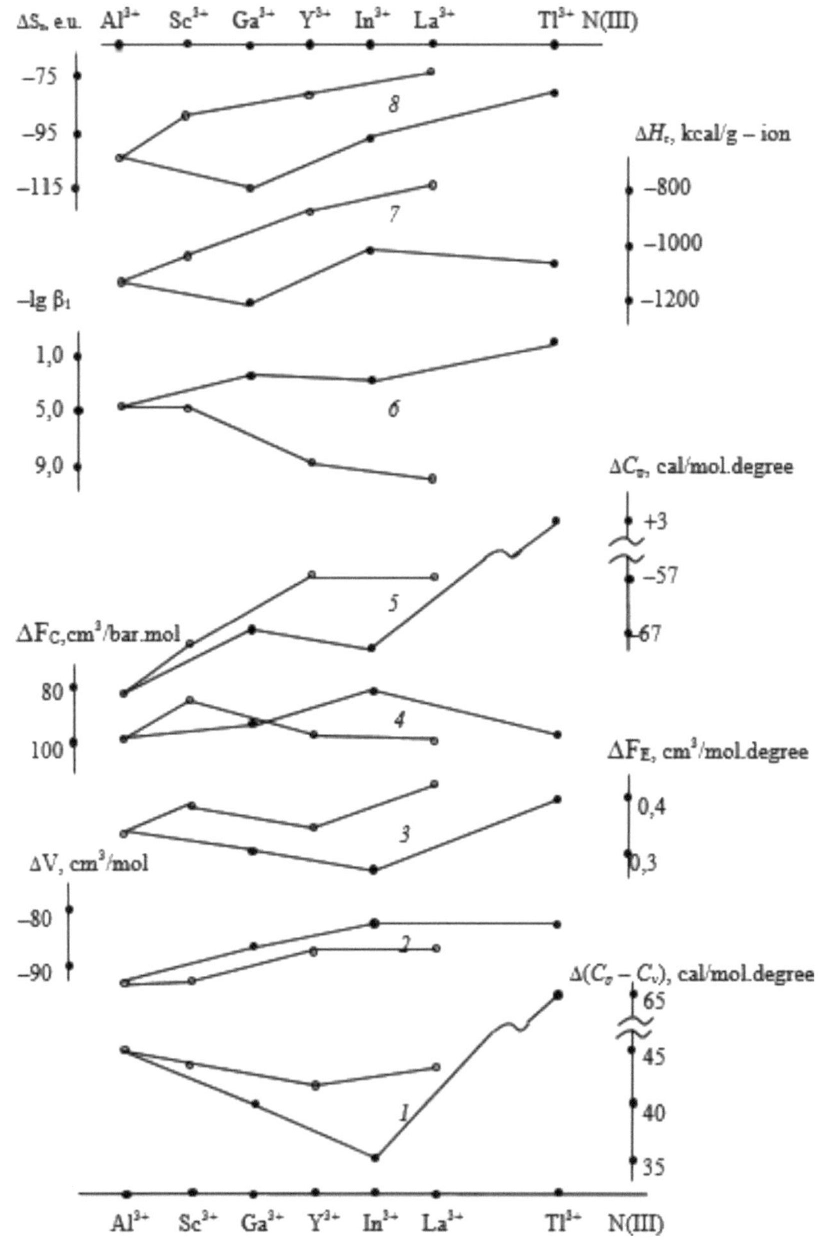

Figure 4.
Dependence of changes in macro properties during the formation of perchlorate solutions on the hydration characteristics of group III cations: 1 - $\Delta(C_p - C_v)$ (1.0 M solutions) (C_p and C_v isobaric and isochoric heat capacities); 2 - ΔV (specific volume); 3 - ΔF_E (expandability); 4 - ΔF_C (compressibility); 5 - ΔC_r (0.5 M); 6 - log β1 (β1 is the constant of the 1st stage of cation hydrolysis); 7 - ΔHr (enthalpy of hydration of the cation); 8 - ΔSr (entropy of cation hydration) [51].

to subgroup, as well as within the subgroup, changes. For example, in **Figure 5** dependences "ΔV - N_{III}" for the subgroup of gallium and scandium change places with increasing salt concentration. This is probably because at low concentrations the ΔV values reflect largely the change in the volume of the outer-sphere water, and at high concentrations, the water in the inner coordination sphere of cations.

Comparison of the tendencies of changes in the bulk properties of solutions from concentration **Figure 5**, apparently, show that the relative compression of the

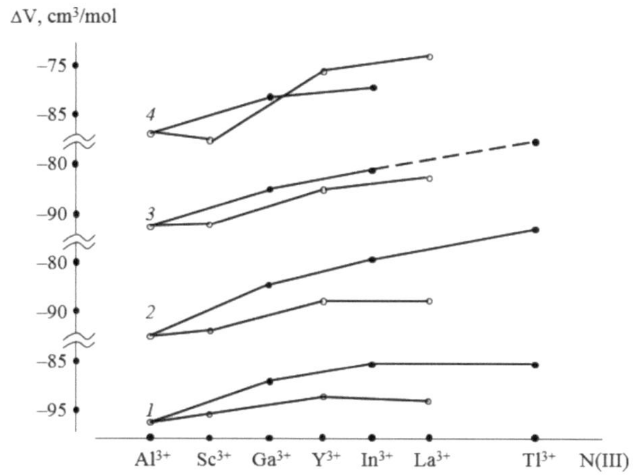

Figure 5.
Dependence of changes in specific volume during the formation of solutions perchlorates with concentration, mol/dm³: 1–0.1; 2–0.5; 3–1.0; 4–3.0 by nature trivalent cations [51].

outer-sphere water is greater for d-metal ions and the compression of water in the inner sphere is, on average, greater for p-metal ions.

In concentrated solutions, the regularities of changes in properties in both groups depend mainly on the state of water in the inner sphere: in the subgroup of d-elements - on the degree of its compression, and in the subgroup of p-elements - on the degree of attraction and polarization. In dilute solutions, an important role is played by the state of the outer-sphere water, namely the degree of its destruction or, conversely, structuring. However, if, in the p-subgroup, the structuring of the outer-sphere water is more typical, and for the d-subgroup, destructuring is most likely [51].

The viscosity and density of perchlorate solutions are significantly influenced by the ability of cations to hydrate. The increased relative kinematic viscosity and the relative activation energy of the aqueous flow of solutions of aluminum, gallium and indium perchlorates is obviously a consequence of the formation of a hydration shell with a branched network of very strong hydrogen bonds by the cation. Within the gallium subgroup, the positive hydration of the ion is expressed the weaker, the larger its radius [52].

In this case, although the anion is not an indifferent component, its nature does not play a particularly important role in the nature of changes in the macro-properties of solutions, since the perchlorate ion has an approximately equal effect on the destruction of the primary structure of water in all systems. The only exceptions are $Tl(ClO_4)_3$ solutions, where the formation of outer-sphere ionic associates with hydrated thallium (III) cations is evident [51].

From the analysis of the compensation spectra of the multiply disturbed total internal reflection, it follows that according to their destructive effect, the anions are arranged in a row $ClO_4^- > NO_3^- >> Cl^-$, and for sulfocation it is not found at all. The formation of a covalent bond “water cation” leads to the strengthening of H-bonds in water, and in the case of solutions of triple-charged cations, additional stabilization of the water structure takes place, which complicates the process of destruction of hydrogen bonds even with such a strongly destructive anion as perchlorate [16].

The interrelation between the formation of outer-sphere water associates and the phenomenon of hydrolysis is especially evident from the comparison of the heat capacities of perchlorates in the series of triply charged cations of metals of the gallium subgroup with the values: 6 - lg β_1 (β_1 is the constant of the 1st stage of hydrolysis of the cation); 7 ΔHr (enthalpy of hydration of the cation); 8 - ΔS_r (entropy of cation hydration) [51, 53, 54] where the polarizing effect of cations on water makes the greatest contribution to the heat capacity even in dilute solutions. All this, moreover, agrees with the results of studying the vibrational spectra of solutions of gallium and indium perchlorates, showing that the water molecules surrounding these ions are polarized not only in the first, but also in the second coordination layer, thus the fraction of polarized water molecules in these solutions is unusual. But high, their contribution to the heat capacity of the solution is also significant [55]. Thallium (III) perchlorate solutions have the highest heat capacity, but here it should be noted that the reason for this in this case may be not only water polarization and local hydrolysis, but also the tendency of Tl^{3+} ions to form perchlorate complexes [56].

The form of the dependences of the apparent molar heat capacity of the $P_{S\ P2}$ of group III metal perchlorates on the salt concentration deviates from the rectilinear, especially for $La(ClO_4)_3$ and $Tl(ClO_4)_3$ **Figure 6a**. The regularities of changes in $P_{S\ P2}$ are also explained in the light of ideas about the energy of interaction of ions with water, the state of water in hydration shells, and the effect of ions on the structure of water. In the gallium subgroup, the role of another factor, the hydrolysis of aquo-cations, increases significantly. Probably, the point for $Ga(ClO_4)_3$ owes its higher position to hydrolysis as compared with indium in **Figure 6b** [49].

Of the entire sum of contributions to the total change in the heat capacity of dissolution of perchlorates, the greatest effect of the specific nature of cations on the heat capacity of solutions occurs through the outer-sphere interactions (ΔC_V). As you can see (**Figure 6b**). for the considered types of electrolytes in a significant range of concentrations ΔC are negative, which may be due not only to the outer-sphere hydration of ions, but also to their destructive effect on a certain part of the outer-sphere water. It is not yet possible to separate these two effects. However, at high concentrations, the sign of ΔC_V hanges to the opposite, indicating, apparently, an increase in the role of the effect of outer-sphere aqueous, as well as ion-ion associates ([50], p. 266).

The recorded changes in the bulk properties of perchlorates of the aluminum subgroup from their concentration are also associated with the individual nature

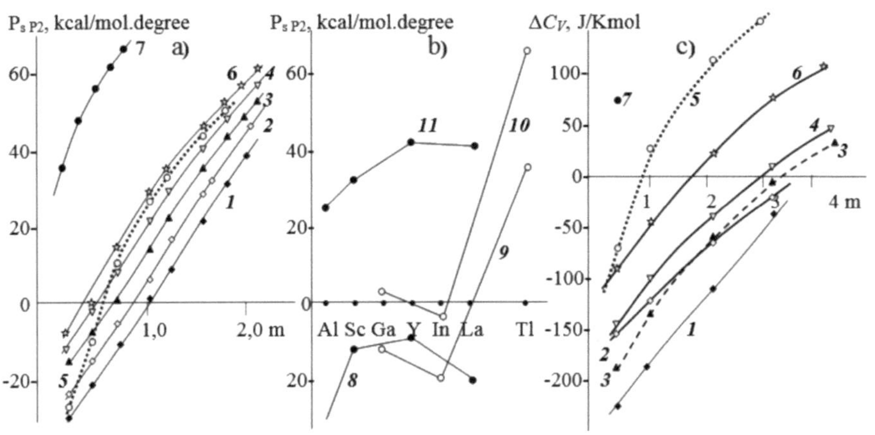

Figure 6.
Dependence of the apparent molar heat capacity of metal perchlorates (P_s) (a, b) [49] and the heat capacity due to outer-sphere interactions of ions (ΔC) (c) ([53], p. 266), on the concentration: 1 - $Al(ClO_4)_3$, 2 - $In(ClO_4)_3$, 3 - $Sc(ClO_4)_3$, 4 - $Ga(ClO_4)_3$, 5 - $La(ClO_4)_3$, 6 - $Y(ClO_4)_3$, 7 - $Tl(ClO_4)_3$, 8 and 9 $Me^{III}(ClO_4)_3$–0.1 m, 10–0.5 m, 11–3.0 m.

of the interaction of ions with water and among themselves. The densities of $Al(ClO_4)_3$, $Ga(ClO_4)_3$ $In(ClO_4)_3$ solutions at temperatures of 24, 25, and 27° C, the speed of sound, specific adiabatic compressibility and isobaric expandability in the concentration range from 0.1 m to saturation, and the shape dependences of apparent molar volumes (F_V), compressibility (F_K) and expandability (F_E).

As follows from **Figure 7** the values of F_V and compressibility F_K monotonically increase with an increase in the electrolyte concentration, but the F_V for $Al(ClO_4)_3$ - decrease, and in the case of $Ga(ClO_4)_3$ and $In(ClO_4)_3$, in all likelihood, pass through a maximum. The dependences of F_V on $m^{½}$ are straightforward, and F_K are curvilinear and for different cations they differ from each other both in their position and in the angle of inclination. The profile of the expansibility curves of PU differs greatly from the characteristics of volumes (F_V) and compressibility (F_K), including the sequence of arrangement of metal ions [49, 50].

The influence of the coordination number on the compressibility of cations is, of course, due to the fact that the coordination water in the first layer practically loses its compressibility. The fact of compressibility manifests itself rather through external-sphere water than through coordination water. This is probably why the hydration numbers of cations, calculated from the compressibility data, turn out to be higher than their c.n. The data are reflected in the **Table 1**, which also shows the change in the isochoric heat capacity of similar electrolytes [53].

The correlation between the polarizing forces of the cation (P) and the eutectic concentration of aqueous solutions of perchlorates (**Figure 8**) indicates that the higher the value of the polarizing force, the stronger the cation interacts with water molecules. Therefore, due to the greater polarizing force of the cations of the gallium subgroup, which characterizes the state of the "cation - water" bond, there is a significant destruction of the intrinsic structure of water and an increase in the number of water molecules drawn from the solution into the sphere of influence of the cation [13].

In addition to indium perchlorate salts, for elements of the aluminum subgroup, their iodates $Me^{III}(JO_3)_3{\cdot}nH_2O$, are also known, including indium iodate with the composition $In(JO_3)_3{\cdot}2H_2O$ [57, 58]. In contrast to perchlorate, indium iodate is characterized by a low degree of solubility, which in water at a temperature of 25° C is ~0,05% by weight (logSP = 2.5), and slightly more passes into solution under the action of nitric acid, showing the maximum value of solubility equal to 16.9 g/dm^3 at a concentration of 300 g/dm^3 HNO_3.

Indium (III) is quantitatively precipitated from the solution in the pH range 1.9–7 with sodium (potassium) periodate with the formation of a precipitate

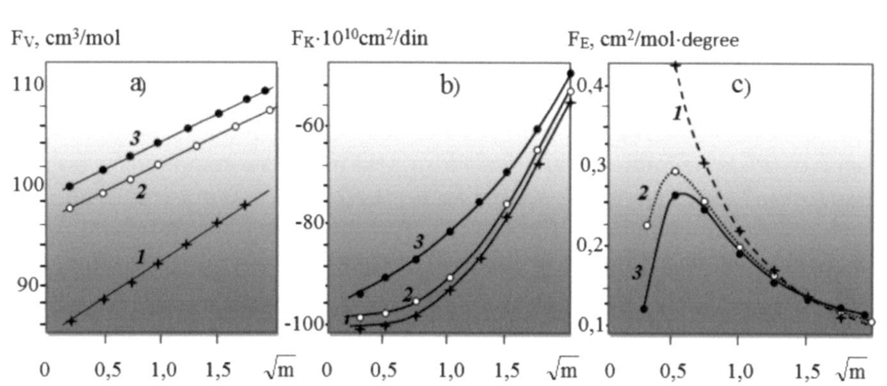

Figure 7.
Dependence of: a – apparent molar volumes (FV), b – compressibility (FK) and c – expandability (FE). From $\sqrt{m}$ perchlorates: 1 - $Al(ClO_4)_3$, 2 - $Ga(ClO_4)_3$ 3– $In(ClO_4)_3$ [49].

Electrolyte	Coordination number of cations	"Acoustic" hydration number	Change in isochoric heat capacity $-\Delta C_{V2}$ (J/K·mol) upon dissolution of subgroup III metal perchlorates on concentration (m)				
			0,5 m	1 m	2 m	3 m	4 m
$Al(ClO_4)_3$	6	12,4	411	373	296	223	
$Sc(ClO_4)_3$,	6	11,3	383	327	248	195	151
$Ga(ClO_4)_3$	7–8	12,5	343	290	233	186	144 (3,83 m)
$Y(ClO_4)_3$	6	13,2	320	276	211	163	114
$In(ClO_4)_3$	6	11,0	345	322	257	205 (3,19 m)	
$Tl(ClO_4)_3$	6	13,4	116				

Table 1.
Change in isochoric heat capacity.

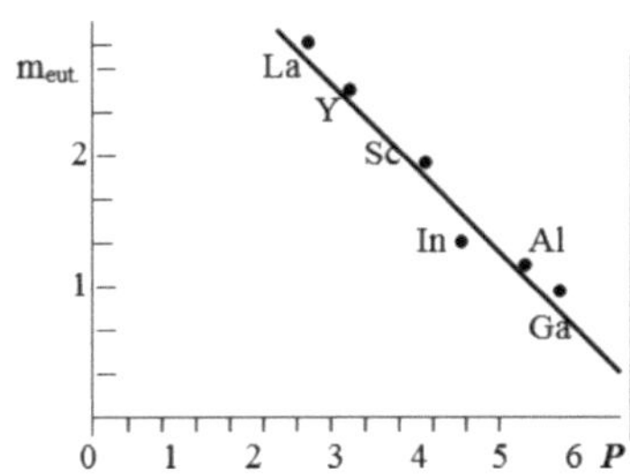

Cation metal	P	$m_{eut.}$	The nature of the connection «$Me^{3+} - H_2O$»	$-\Delta H_r^o$, kJ/mol
Al(IIII)	5,547	1,09	Covalent	4788
Ga(III)	5,796	0,92		4813
In(III)	4,338	1,3		4240
Y(III)	3,093	2,26	Weakly covalent	3750
La(III)	2,79	3,02		3411

Figure 8.
Correlation of the polarizing forces of cations (P) and the eutectic concentration (m_{eut}) of aqueous solutions of group III perchlorates $Me^{III}(ClO_4)_3$ ([50], pp. 124–126).

corresponding to formula $In_5(JO_6)_3$, the composition of which is constant at a temperature of 120–270° C. The crystalline compound of indium paraperiodate is slightly soluble in water, and respectively, in 1%, 0.3% and 0.5–1% carbonate, nitrate and ammonium acetate, but it is readily soluble in dilute mineral acids. The solubility of indium paraperiodate in water at 20° C is $6{,}4\cdot10^{-4}$ g-mol/dm^3 or $7{,}9\cdot10^{-3}$ g/dm^3. The gravimetric and titrimetric methods for the determination of indium using the periodate ion are based on these analytical characteristics [59].

Answering the question whether the ClO_4^- ion is included in the inner coordination sphere of indium, it can be considered proven that indium acidic complexes are not formed in aqueous solutions, and in this case we are only talking about the possibility of the outer-sphere interaction of ClO_4^- ions with aquocation. However, in a number of organic systems, apparently, their participation in complex formation is possible, but such states, being weaker than contact interactions, are rather difficult to fix. Perchlorate ion and perchloric acid in indium compounds perform primarily the function of a dehydrating agent; therefore, their presence leads to a change in the characteristics of the interaction of ions with water, which is reflected in the physicochemical properties of indium perchlorates.

Apparently, in relatively dilute solutions, the role of ClO_4^- is reduced to destructuring water, which contributes to the stabilization of hydration shells around hydrophilic indium cations. In concentrated solutions, ClO_4^- anions enter into some competition with cations for water, as a result of which, in particular, an increase in the asymmetry of coordination water molecules and its additional polarization is possible. Indium (III) cations attach a larger number of layers of water and, to varying degrees, affect its state, and through it, the action spreads further to the property and the boundary medium. Thus, this causes either the destructuring effect or a change in the state of chemical bonds of ClO_4^- ions in concentrated solutions.

Thus, the ClO_4^- ion, being structure-forming and exhibiting a protoacceptor ability, in solutions of group III perchlorates forms exclusively solvate-separated ion pairs due to the high enthalpy of hydration of the corresponding metal cations.

Author details

Boris Radionov Konstantinovich and Ilya Svirsky Anatolievich*
Ural Federal University Named After B.N. Yeltsin, Yekaterinburg, Russia

*Address all correspondence to: sviskill.171993@gmail.com

References

[1] Johansson L. The role of the perchlorate ion as ligand in solution// Coordinat. Chem. Revs. 1974. V.12. № 3. P.241-261.

[2] Rosolovsky V.Ya. Complex perchlorates of polyvalent metals. XI Mendeleev Congress on General and Applied Chemistry. Inorganic chemistry and technology of inorganic substances. The science. M.: 1975.S. 82-83.

[3] Cozzi D., Vivarelli S. Beitrag zur Chemie des Indiums. // Z. Elektrochem. 1953. Bd. 57. № 16. S. 408-416.

[4] White J.M., Kelly P., Li N.C. Dynonyl naphthaline sulphonic acid and tri-n-octylamine as liquid ion-exchangers for the stydy of Fe(III) and In(III) chloride complexes // J. Inorg. Nucl. Chem. 1961. V. 16. № 3-4. P. 337-344.

[5] Hester R.E., Plane R.A. a) Solvation of metal ions in aqueous solutions: the metal-oxygen bond // Inorg. Chem. 1964. V. 3. № 5. P. 768-769;b) A Raman spectrophotometric comparison of interionic association in aqueous solutions of metal nitrates, sulfates, and perchlorates // Inorgan.Chem. 1964.V. 3. № 5. P.769-770.

[6] Hester R.E., Plane R.A., Walrafen G.E. Raman spectra of aqueous solutions of indium sulfate, nitrate, and perchlorate // J. Chem. Phys. 1963. V. 38. № 1.P. 249-250.

[7] Hester R.E., Plane R.A. A Raman spectrophotometric comparison of interionic accociations in aqueous solution of metal nitrates, sulfates, and perchlorates // Inorg. Chem. 1964. V. 3. № 5. P. 769-770.

[8] Hester R.E., Grossman W.E.L. Low frequency bands in the Raman spectra of aqueous indium (III) solutions // Spectrochim. Acta. 1967. V. 23A. № 6.P. 1945-1946.

[9] Rudolph W.W., Fischer D., Tomneg M.R., Rye C. C. Indium (III) hydration in aqueous of perchlorate, nitrate and sulfate. Raman and infrared spectroscopic studies and ab-initio molecular orbital calculation of indium (III) – water cluster // Phys. Chem.Chem.Phys. 2004. V. 6. № 22. P. 5145-5155.

[10] L.A. Mund Investigation of aqueous solutions of metal perchlorates of the first and third groups of the periodic system of elements by vibrational spectroscopy / Author's abstract. dissertation. for a job. learned. step. kand. chem. sciences. L. 1973.25 C. (Leningrad State University named after A.A. Zhdanov).

[11] Tuck D.G. The coordination chemistry of In (III) isreviewed // Coord. Chem. Rev. 1966. № 1. P. 286-291.

[12] Latysheva V.A., Myund L.A., Glebovskiy D.N. Protolithic interaction of water with aquacations of some metals according to infrared spectroscopy // J. physical Chemistry. 1972. T. 46. No. 5.S. 1330-1332.

[13] Davidyan A.G. The structure of aqueous solutions of metal perchlorates of I-III groups of the Periodic system D.I. Mendeleev / Dissertation for the degree of Candidate of Chemical Sciences San Petersburg. 2014. S. 153.

[14] Davidyan A.G., Kudrev A.G., Myund L.A., Khripun M.K. The structure of aqueous solutions of group III metal perchlorates according to NIR spectroscopy // J. general chemistry. 2013. T. 83. No. 3. S. 359-367.

[15] Davidian A.G., Kudrev A.G., Mynd L.A., Khripun M.K. Near infrared spectral studies of aqueous solutions of matal perchlorates in group I A, II A, II B, III A and III B of the periodic table //

J. Near Infrared Spectralosc.2014. V. 22. № 1. P. 27-34.

[16] Kozhevnikova G.V., Mikhailov B.A., Burkov K.A. Investigation of the influence of some ions on the state of water in electrolyte solutions using the compensation method of spectrophotometry MNPVO // In sb. "Molecular physics of biophysics of water systems". L.: LSU. 1973. No. 1. S. 79-83.

[17] Cannon T.H., Richards R.E. Magnetic resonance studies of solution containing ^{115}In // Trans. Farady. Soc. 1966. V. 62. № 6. P. 1378-1387.

[18] Buslaev Yu.A., Tarasova V.P., Buslaeva M.N., Petrosyants S.P. Investigation of aqueous solutions of electrolytes by NMR on nuclei of cations // Dokl. Academy of Sciences of the USSR. 1973. T. 209. No. 4. C. 882-884.

[19] Šmirous F., Čeleda J., Palek M. Contributions to the chemistry of highly concentrated aqueous electrolyte solutions. XXV. Ionophoretic investigation of the association of Al^{3+}, Sc^{3+}, Ga^{3+} and In^{3+} ions in concentrated solutions of $NaClO_4$ and LiCI // Collect. Czech. Chem. Communs. 1971. V. 36. № 12. P. 3891-3899.

[20] . Sakellaridis P., Coromanzou M. Etude spectrophotométrique des solutions chlorhydriques de l'indium // Bull. Soc. chim. France. 1962. № 4. S. 769-771.

[21] Schnoor R. The influence of liquid junction potentials on potentiometric determination of the activity coefficient of In^{3+} ion in concentrate aqueous perchlorate solutions // Z. Phys. Chem. 1989. Bd. 270. № 6. S. 1169-1176.

[22] Fratiello A., Vidulich G.A., Cheng C., Kubo V. A direct hydrogen-1 andphosphorus-31 nuclear magnetic resonance cation solvation study of $Al(ClO_4)_3$, $Ga(ClO_4)_3$, $In(ClO_4)_3$, $UO_2(ClO_4)_2$ and $UO_2(NO_3)_2$ in water-acetone-dimethylsulfoxide and water-acetone-hexamethylphosphor-amide mixtures // J.Solut. Chem. 1972. V. 1. № 5. P. 433-444.

[23] Fratiello A., Lee R.E., Nishida V.M., Schuster R.E. Proton magnetic resonance coordination number study of Al (III), Be (II), Ga (III), In (III) and Mg (II) in water and aqueous solvent mixtures // J. Chem. Phys. 1968. V. 48. № 8. P. 3705-3711.

[24] Lincoln S.F. Solvent coopdination numbers of metal ions in solution // Coopd. Chem. Rev. 1971. V. 6. № 4. August. P. 309-329.

[25] Golub A.M., Samoilenko V.M. Influence of non-aqueous solvents on the formation of indium (III) thiocyanate complexes // Ukr. chem. zhurn. 1963. T. 29. No. 6. C. 590-600.

[26] Levin I.S., Rodina T.F., Yukhin Yu.M. et al. Extraction of elements of the subgroup gallium and arsenic from solutions of oxygen-containing acids with di- (2ethylhexyl) phosphoric acid / In collection of articles. "Chemistry of extraction processes". M .: Science. 1972.S. 262-267.

[27] Diamond R.M. The solvent extraction behavior of inorganic compounds. IV. Variation of the distribution quotient with chloride, hydrogen ion held constants // J. Phys. Chem. 1957. V. 61. № 11. P. 1522-1531.

[28] Hasegawa Y., Takeuchi H., Sekine T. The behavior of indium(III) complexes in several liquib-liquib distribution system. 2. The extraction of indium (III) halide and thiocyanate complexes into 4-methyl-2-pentanone, 4-methyl-2-pentanol, or nitrobenzene // Bull. Chem. Soc. Japan. 1972. V. 45. № 5.P. 1388-1394.

[29] Brunette J.P., Taneri M., Goetz-Grandmont G., Leroy M.J.F.a) Synergic extractions of indium (III)

from chloride and perchlorate media 1-phenyl-3-methyl-4-acylpyrazol-5-ones and tri-*n*-octylphosphophine oxide mixtures in toluene // Solv. Extr. and Ion Exch. 1985. V. 3. № 3. P. 309-330;b) Extractions of indium (III) from chloride media with 1-phenyl-3-methyl-4-acylpyrazol-5-ones. Synergic effect with high molecular weight ammonium salts // Polyhedron. 1982. V. 1. № 5. P. 457-460.

[30] Genge J.A.R., Salmon J.C. Ion-exchange studies of phosphates. Part III. Complex formation between tervalent metals and orthophosphoric acid // J.Chem. Soc. 1959. Apr. P. 1459-1463.

[31] Gorokhova L.N., Tsintsevich E.P. Cation exchange of gallium and indium in the presence of perchloric acid and organic solvent // Moscow State University Bulletin. 1966. No. 3. P.113-114.

[32] Radionov B.K., Maltsev G.I. Indium in aqueous solutions. Chemistry of rare and trace elements. // Germany, Saarbrücker: LAP LAMBERT Academic Publishig. 2014.-352 p.

[33] Golub A.M., Akmyradov R., Uskova S.L. a) Study of indium nitrite complexes in methanol // Izv. Universities. Chemistry and chem. technology. 1974. T. 17. No. 9. S. 1287-1290; b) Coordination compounds of indium nitrite with some organic ligands // J. inorg. chemistry. 1974. T. 19.No. 7.P.1761-1764.

[34] Ionov V.I., Chicherina N.Yu. Extraction of indium tenoyltrifluoro-acetate from perchlorate and chloride solutions // J. inorg. chemistry. 1979 T. 24. No. 2. S. 465-470.

[35] Gusev N.I. Study of ion hydration by the method of electrical conductivity a) XIII. Some issues of ion hydration and the structure of solutions // J. physical chemistry. 1973. T. 47. No. 9. S. 2327-2331; b) IX. Association in acidic perchlorate solutions // J. physical Chemistry. 1973. T. 47. No. 2. S. 324-328; c) X. Change in the number of water molecules oriented by cations in the temperature range from 10 to 50 ° C // Ibid. 1973. T. 47. No. 1. S. 96-100.

[36] Pestova O.N., Davidyan A.G., Myund L.A., Khripun M.K. Solubility of perchlorates of aluminum, gallium, indium in water // J. general chemistry. 2011. T. 81.No. 8.S. 1237-1241.

[37] Heinje G., Luck W.A.P., Heinzinger K. Molecular dynamics simulation of an aqueous $NaClO_4$ solution // J. Phys. Chem. 1987. V. 91. № 2. P.331-338.

[38] Paskalev N. Basis perchlorates of indium (III) // Dokl. Bolg. AN. 1967. T.20. № 6. C. 549-552.

[39] Celeda J., Tuck D.G. A densimetrie study of indium (III) species in meneral acid solutions // J. Inorg. and Nucl. Chem. 1974. V. 36. № 2. P. 373-378.

[40] Maeda M., Ohtaki H. An X-ray diffraction study on the structure of the aqua indium (III) ion the perchlorate solution // Bull. Chem. Soc. Jap. 1977. V. 59. № 7. P. 1893-1894.

[41] Ilyukhin A.B., Malyarik M.A. The crystal structure of indium perchlorates $[In(H_2O)_6](ClO_4)_3$ and $[In(H_2O)_6](ClO_4)_3 \cdot 3H_2O$ // J. inorg. Chemistry. 1999. T. 44.No 4.S. 532-535.

[42] Seward T. M., Henderson C. M., Charnock J. M. Indium (III) chloride comphlexing and solvation in hydrothermal solutions to 350°C: an EXAFS study // Chem. Geol. 2000. V. 167. № 1-2. P. 117-127.

[43] Golub A.M., Samoilenko V.M. Potentiometric study of the composition and strength of ion solvates, J. inorg. chemistry ,. 1965. T. 10. No. 2. S. 328-331.

[44] Haladjian J., Bianco P. Préparation des perchlorates d'aluminium, de gallium et d'indium anhydres en solution méthanolique // Chemia. 1973. V. 27. № 11. P. 588-589

[45] Petrů F., Kůtek F.Ü ber die thermische Zersetzung von Perchlorates einiger Metalle der III. Gruppe // Z. Chem. 1964. Hf. 1. Jg 4. S. 33-34.

[46] Carty A.J., Tuck D.G. Co-ordination compounds of indium. Part.I. Some cationic complexes of indium (III) // J. Chem. Soc. 1964. Suppl. № 2. P. 6012-6017.

[47] Ishihara K., Funahashi S., Tanaka M. Mechanistic study on the complex formation of aluminum (III), gallium (III), and indium (III) ions in dimethyl sulfoxide and N,N-dimethylformamide // Inorgan. Chem. 1986. V. 25. № 16. P. 2898-2901.

[48] Cherkasova T.G., Tatarinova E.S. Crystal structure of indium hexa (dimethyl sulfoxide) perchlorate (III) // Izv. VUZov. Chemistry and chemical technology // 1997. V. 40. No. 2. P. 26-28.

[49] Babakulov N., Latysheva V.A.a) Bulk properties of aqueous solutions of aluminum, gallium and indium perchlorates // Izv. Universities. Chemistry and chemical technology. 1973. T. 16. No. 9. S. 1449-1451;b) Heat capacity of aqueous solutions of metal perchlorates of the third group of the periodic system // J. physical chemistry. 1974. Vol. 48. No. 4. P.1012-1014.

[50] Babakulov N., Grigorieva E.G., Latysheva V.A. The speed of sound and the compressibility of aqueous solutions of perchlorates of some metals of the second and third groups of the Periodic Table D.I. Mendeleev. // Rukop. dep. VINITI. 1976. No. 3004-76 Dep.

[51] Babakulov N., Latysheva V.A., Shchukarev S.A. The nature of p- and d-metal cations of the third group of the periodic system and the properties of aqueous solutions of their perchlorates // J. general chemistry. 1977. Vol. 47, No. 1.S. 23-31.

[52] Penkina N.V., Smaev V.N. Viscosity and density of aqueous solutions of perchlorate cations of group III. II. Gallium and indium perchlorates. // Rukop. dep. VINITI. 10/04/74. No. 2597-74 Dep.

[53] Latysheva V.A. Water-salt solutions. Systems approach. SPb .: Publishing house. St. Petersburg. un-that. 1998. 344 S.

[54] Latysheva V.A. Heat capacity of solutions and symmetry of aqua ions // Journal of Physics. chemistry. 1974. T. 48. No. 5. S. 1315. Rukop. dep. VINITI. 04.02.74. No. 213-74 Dep.

[55] Latysheva V.A., Myund L.A., Glebovskiy N.D. The state of aqua ions in aqueous solutions of metal perchlorates of 1-3 groups of the periodic system of metals according to vibrational spectroscopy / In collection. "Problems of modern chemistry of coordination compounds" L .: LGU, 1975. Issue 5.S. 85-121.

[56] Rogers T.E., Waind G.M. Spectrophotometric studies of concentrated perchlorate solutions. Part. I. Thallic perchlorat // Trans. Far. Soc. 1961. V. № 8. P. 1360-1370.

[57] Mathers F.C., Schluederberg C.G. Obtaining perchlorate, iodate, selenate and cesium-selenium indium alum // J. Amer. Chem. Soc. 1908. V. 30. P. 211-215.

[58] Ivanov-Emin B.N., Rybina V.I., Zaitsev B.E. et al. a) Study of aluminum iodates and elements of the gallium subgroup // Coord. chemistry. 1979. T. 5. No. 4. S. 514-519;b) Investigation of the thermal stability of aluminum iodates and elements of the gallium subgroup //

J. inorg. chemistry. 1981. T. 26. No. 12. S. 3227-3230.

[59] Vertizade A.A., Mekhtiev M.M. Gravimetric and titrimetric methods for the determination of indium using the periodate ion // Zh. analyte. chemistry. 1970. T. 25. No. 6. S. 1097-1101.

Chapter 4

Investigation of Indium Oxide Effect on Indium Particles Properties Used as Silicon Nanowires Catalyst

Rabia Benabderrahmane Zaghouani

Abstract

In this chapter, we investigate on indium particles elaboration by different annealing processes: rapid thermal annealing (RTA) and conventional processes. The elaborated particles are dedicated to be used as catalyst for silicon nanowires' (SiNWs) growth by vapor–liquid–solid (VLS) process. The annealing parameters effect on indium particles properties is studied. After conventional annealing, the indium layer is cracked into elongated and inhomogeneous islands of micrometric sizes. XRD analysis depicts, in addition to pure indium planes, the presence of new peaks attributed to indium oxide (In_2O_3) planes formed during annealing. After hydrogen treatment with a flow rate of 60 sccm during 10 min, some In_2O_3 peaks are eliminated and replaced by new indium peaks, explaining the amelioration of indium particles morphology. These formed particles have been used as catalyst for SiNWs' growth. A low density of SiNWs is obtained, attributed to In_2O_3 persistence, decreasing the indium catalytic effect. Quasi-spherical and homogeneously distributed indium particles with an average size of 422 nm are successfully grown in one step by the RTA process during short time (5 min) at lower temperature (450°C). XRD analysis shows the absence of indium oxide in the contrary to those formed by the conventional furnace. SiNWs with higher density are obtained, highlighting the harmful role of indium oxide.

Keywords: indium catalyst, annealing, indium oxide, silicon nanowires, vapor–liquid–solid process

1. Introduction

The advancement of nanotechnologies has made possible the development of new applications in all fields. In particular, the nanostructuring of materials has paved the way toward new concepts. Important efforts have been dedicated to metallic nanoparticles, thanks to their interesting properties in comparison to bulk materials. Specially, they are widely studied to be used for many purposes in various applications: passivation of silicon for photovoltaics [1], plasmonics [2, 3], and bionanotechnology [4]. Metallic nanoparticles are also used as catalyst for silicon nanowires' (SiNWs) growth by top-down [5, 6] or bottom-up approaches [7].

Silicon nanowires' properties, especially their diameter, could be tuned by controlling the metallic particles properties. In the literature, many metallic nanoparticles are reported as catalysts, such as Au [8], Ti [9], Al [10], Cu [11], Ga [12], and Pt [13]. During last years, indium is considered a very interesting catalyst because it forms a low temperature eutectic with silicon (157°C) and it induces shallow defects as it acts as a p-type dopant encouraging its use as catalyst. An important issue in SiNWs' synthesis is the indium catalyst elaboration.

Indium nanoparticles could be elaborated by different techniques such as vapor deposition technique [14], electrochemical reduction [15], chemical reduction of salts [16], laser ablation [17], and reduction of indium-tin oxide or indium layers by hydrogen or helium plasma [18–20]. Iacopi et al. have obtained indium particles with diameter range of 40–80 nm by electrodeposition on silicon substrate from an aqueous solution ($InCl_3$, KCl, and HCl) [21]. Kumar et al. have reported on the growth of indium droplets obtained with an average diameter of 90 nm by indium evaporation followed by annealing at 300°C during 5 min [22]. The obtained particles' properties are closely related to experimental parameters such as the precursor's concentration, the plasma flow rate, the exposition duration and the substrate temperature, or the annealing parameters.

In this chapter, we report on an ex situ formation of indium particles to be used as catalyst for SiNWs' growth using two annealing processes: a rapid thermal annealing (RTA) and a conventional process. A comparative study is carried out to investigate the annealing process effect on SiNWs' properties. In particular, the effect of indium oxide is presented.

2. VLS process

The bottom-up approach is presented as an interesting alternative for low-cost nanowires growth. Indeed, it requires few technological steps with the possibility of SiNWs' growth on any substrate. Several techniques have been reported, mainly chemical vapor deposition (CVD), plasma-assisted chemical vapor deposition (PECVD), laser ablation, and molecular beam epitaxy (MBE). PECVD is the technique adopted in this work. It has the same principle as the conventional CVD except that the chemical reactions will take place after the formation of plasma from the reactor gases, offering the possibility to work at low temperatures. The deposited layers' properties strongly depend on the substrate temperature, the pressure, the growth time, the reactive gases, and the gas flow rates. The principle of SiNWs' growth by PECVD can be resumed in four main steps:

1. The deposition of catalytic particles (in situ or ex situ).

2. The formation of a metal-silicon eutectic by supplying the particles with a precursor gas [the silane (SiH_4) in the case of silicon nanowires].

3. Saturation of metal particle with silicon (Si), nucleation and precipitation of silicon at the substrate–metal particle interface.

4. The growth of SiNWs.

The SiNWs' growth is generally explained by the vapor–liquid–solid (VLS) mode proposed by R. S. Wagner and W. C. Ellis. In this mode, the growth depends on three main elements: the precursor in its gaseous state, and the metal-silicon alloy in the liquid state, the nanowire in the solid state, and hence the nomination vapor–liquid–solid mode.

The SiNWs' growth in a PECVD reactor is carried out by following the next steps:

- At $T_{work} > T_{eutectic\ of\ metal\text{-}Si}$: the metallic particle is in its liquid state.
- Introducing the precursor gas (SiH_4): The silane molecules in the vapor state are adsorbed on the particle surface according to the following equation:

$$SiH_4 \rightarrow Si + 2H_2 \tag{1}$$

- Incorporation of Si atoms in the droplet, formation of the metal-Si alloy, and the silicon diffusion toward the alloy-substrate interface. The silicon concentration in the droplet will exceed the equilibrium concentration at the working temperature, leading to the droplet saturation and the silicon nucleation.

3. Indium as catalyst for silicon nanowires' growth

In the VLS mode, the catalyst surface's properties play an important role in the SiNWs' growth. In order to adsorb the maximum of gaseous species, the surface should be rough. For example, silane dissociation and adsorption are better on the gold-silicon (Au-Si) system's surface than that of silicon. Indeed, the high adsorption and dissociation efficiency of silane at the Au-Si droplet permits the nanowires' growth with a constant radius. Other parameters influence SiNWs' growth, such as the surface tension of the catalyst droplet and the solid–liquid interface tension. The choice of the catalyst metal is very important, matching some criteria:

- The eutectic temperature of the metal-silicon alloy
- Vapor pressure
- Silicon solubility in the metal
- Technological compatibility with the current semiconductor industry
- Metal diffusion in SiNWs and formation of recombination centers
- Oxidation, etc.

Gold is the most commonly used metal despite silicon contamination with defects, introducing deep energy levels in the silicon bandgap. Gold, with a simple diagram phase, permits the solubility and nucleation of silicon easily at a relatively low eutectic temperature (363°C). Moreover, gold does not oxidize, increasing its catalytic activity. Gold-catalyzed SiNWs are well controlled and already integrated into prototypes such as transistors [23], biosensors [24], and photo-anodes [25].

Since 2001, post-transition metals like aluminum, indium, and bismuth have been studied. All these metals introduce shallow defects (dopants for silicon). Aluminum, for example, has an eutectic temperature (577°C) higher than that of gold and is very chemically reactive, especially with oxygen.

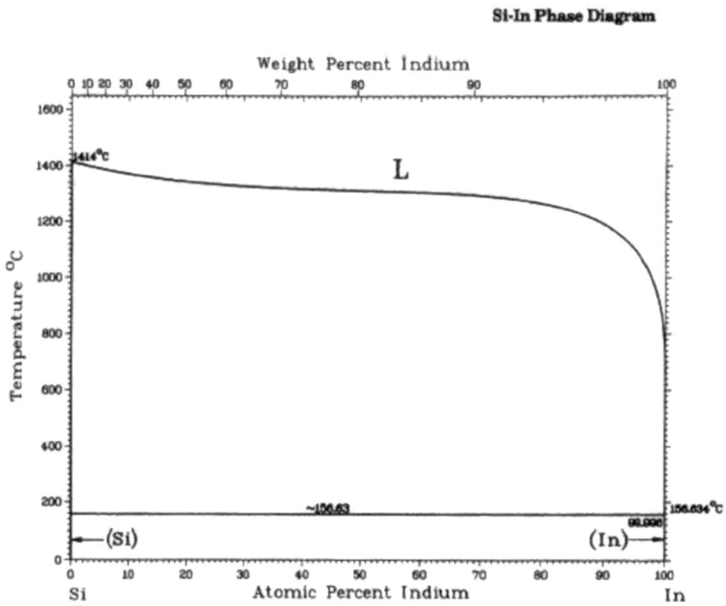

Figure 1.
Phase diagram of indium-silicon system [26].

Bismuth (Bi) is a promising catalyst. The energy level of the impurity introduced by bismuth into the silicon gap is close to the conduction band, so bismuth is an N-type dopant for silicon. Bismuth exhibits a low silicon eutectic temperature (271°C), which allows working at low growth temperatures and using flexible substrates. Despite these advantages, it is difficult to use it as SiNWs catalyst because it has a high vapor pressure, so it can evaporate easily during growth. It has a low surface tension (0.37 N/m), while gold has a higher surface tension (1.14 N/m). Unidirectional growth has been shown to be difficult with low surface tension catalysts.

Indium is an interesting metal to be used as catalyst. It presents a simple phase diagram with silicon (**Figure 1**), forming an eutectic system at 157°C, permitting low growth temperatures and flexible substrates' use. Indium has a surface tension (0.55 N/m) higher than that of bismuth but lower than gold. However, in the presence of oxygen, indium could be oxidized, reducing thereby its catalytic behavior.

4. Experimental details

Figure 2 describes the silicon nanowires growth process [27]. Indium particles (In-Nps) were grown ex situ by annealing indium-coated silicon p-type (100) substrates. Indium layers of 100 nm thickness were deposited on silicon substrates by thermal evaporation. The used annealing procedures are: (1) conventional annealing with a vacuum pressure of 10^{-2} mbar at 600°C during 45 min, (2) RTA annealing using a home-built RTA furnace with a vacuum pressure of 10^{-6} mbar at different temperatures (300, 350, 400, and 450°C) during 5 min.

To synthesize SiNWs, the samples are introduced to the PECVD reactor with substrate temperature set to 400°C. Prior to precursor (SiH_4) introduction, the samples are treated by hydrogen plasma during 10 min with a flow rate of 60 sccm. Then, SiH_4 is introduced during 15 min with a flow rate of 10 sccm.

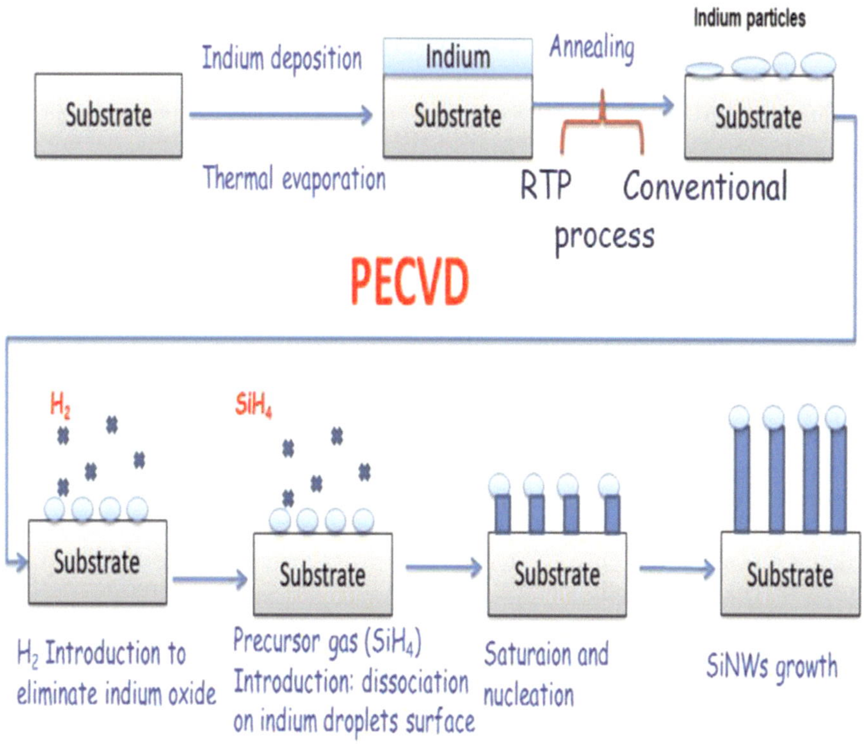

Figure 2.
SiNWs' growth process adopted in this work.

5. Annealing process effect on indium particles' catalyst

The indium-coated silicon substrate annealed in the conventional furnace during 45 min at 600°C shows elongated and inhomogeneous islands of micrometric sizes (**Figure 3**). The chosen work temperature is well above the indium melting temperature (157°C), offering sufficient kinetic energy to metallic atoms to form regular particles. The observed morphology could be attributed to the presence of indium oxide. X-ray diffraction (XRD) analysis was performed to confirm these observations.

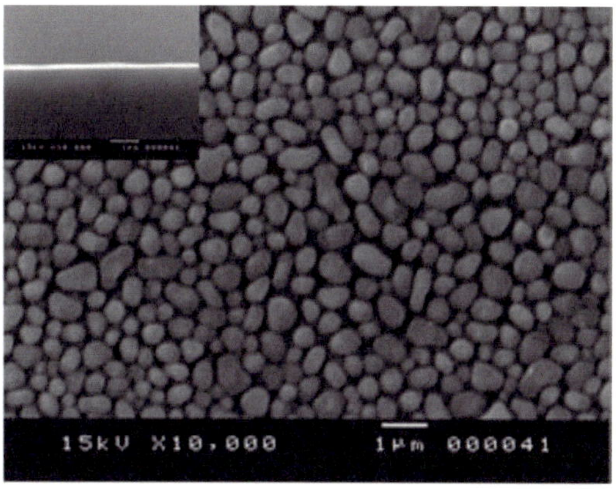

Figure 3.
(a) SEM image of indium-coated silicon substrate annealed under conventional process. Inset: Cross-sectional SEM view of the as-deposited indium-coated silicon substrate.

In **Figure 4**, we compare the XRD spectra of the as-deposited indium layer and the annealed sample. The XRD patterns of the as-deposited sample show only the presence of the planes of the tetragonal crystal structure of indium. However, after annealing, the XRD spectrum shows the disappearance of indium peaks except two peaks. These peaks are replaced by peaks attributed to indium oxide planes with the presence of very intense (222) indium oxide peak. Indium oxide can explain the obtained indium particles morphology. This phenomenon is attributed to the high melting temperature (1900°C) of indium oxide compared to the chosen annealing temperature.

So, in order to ameliorate the indium particles properties, a hydrogen plasma treatment was performed in a PECVD reactor at a substrate temperature of 400°C during 10 min, with two different flow rates of 60 and 100 sccm. XRD analysis shows the persistence of indium oxide presence after 60 sccm H_2 treatment, indicating that this flow rate is not sufficient to eliminate all In_2O_3. When increasing hydrogen flow, In_2O_3 peaks have been disappeared and replaced by indium peaks (**Figure 5**). This result is confirmed by the SEM image of the obtained structures (**Figure 6**). It is noticed that the indium particles become more homogeneous and regular in size and form. Quasi-spherical particles with average size of 440 nm (**Figure 6(c)**) are obtained.

H_2 plasma treatment leads to the indium oxide elimination as depicted by Eq. (2) and the indium loss through evaporation leading to particles properties enhancement (density, size, and shape). Moreover, XRD results highlight the fact that the H_2 plasma flow rate (100 sccm) was sufficient to eliminate all indium oxide.

$$3H_2 + In_2O_3 \rightarrow 2In + 3H_2O \tag{2}$$

5.1 Discussion

In this section, we have noted the indium oxide formation during the conventional annealing that is attributed to the oxygen presence. This observation is explained by the low vacuum pressure in the furnace (10^{-2} mbar) where the heating is mainly provided by thermal conduction. In order to eliminate the oxygen and

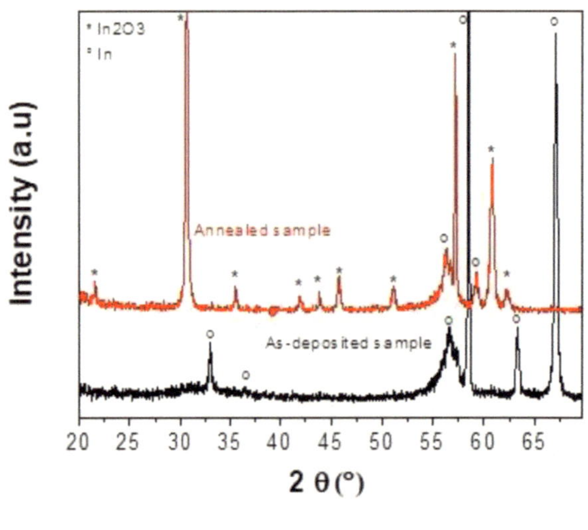

Figure 4.
XRD spectra of as-deposited indium layer and annealed sample.

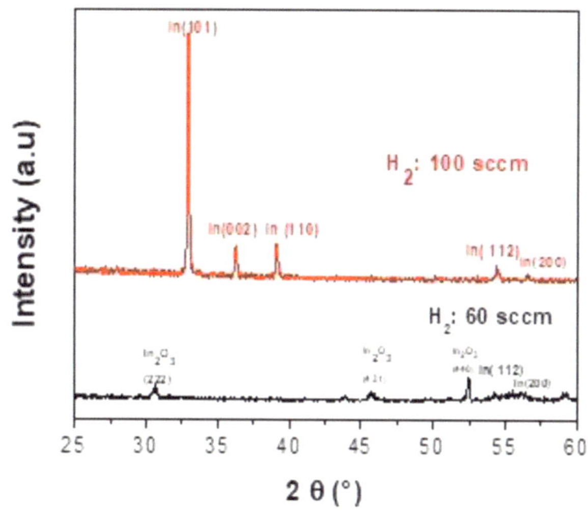

Figure 5.
XRD spectra of annealed layers and treated by H_2 treatment with flow rates of 60 and 100 sccm.

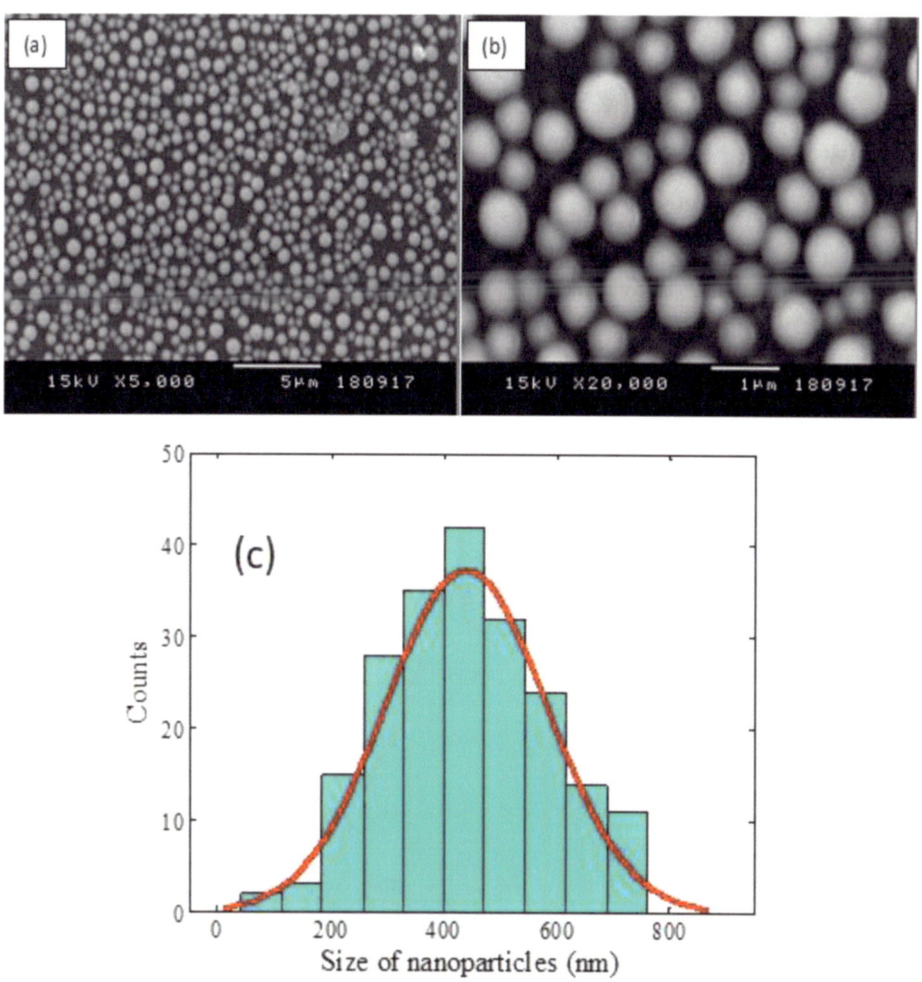

Figure 6.
(a-b) SEM images of indium particles obtained after conventional annealing followed by H2 treatment with flow rate of 100 sccm (c) the corresponding histogram indicating the size distribution.

form pure indium particles, samples must be annealed in ultra-vacuum atmosphere; however, the thermal conduction will be very slow. In this case, the annealing will occur for long durations consuming thereby much energy.

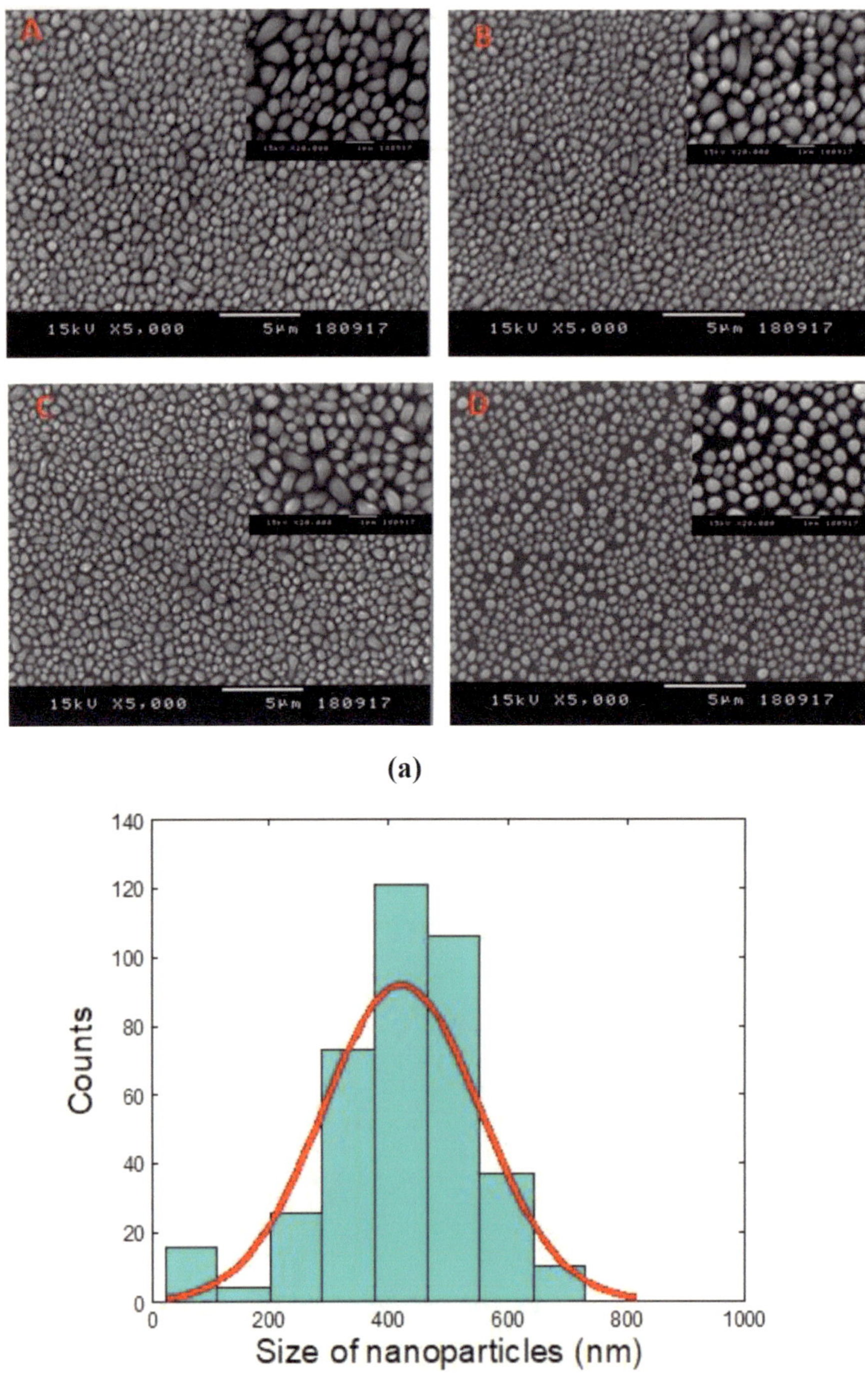

Figure 7.
(a) SEM image of indium particles obtained after RTA annealing; A: 300°C, B: 350°C, C: 400°C, and D: 450°C and (b) the corresponding histogram indicating the size distribution.

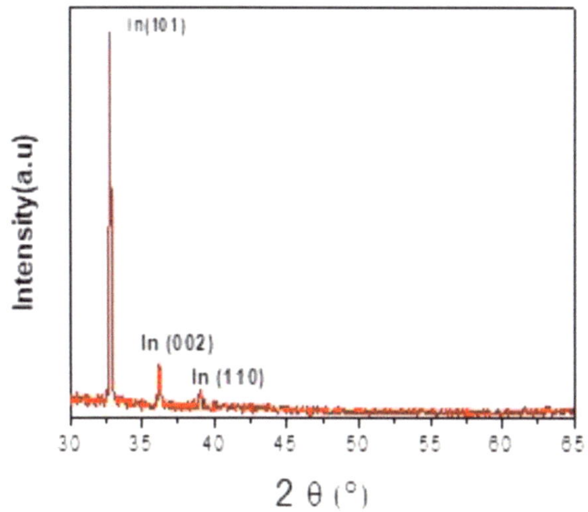

Figure 8.
XRD patterns of the sample annealed at 450°C.

In order to overcome this problem, RTA annealing based on radiation heating by infrared short waves (400–1400 nm) is used. The heating duration could be decreased, thanks to high silicon absorption in this wavelength range.

Indium-coated substrates were annealed at different temperatures (300, 350, 400, and 450°C) during 5 min. It is noticed in **Figure 7(a)** that for the temperatures 300 and 350°C, the surface morphologies are quite similar. The substrates are covered with inhomogeneous particles in size and shape, with a high surface density. An improvement in the particles shape is obtained at 400°C. At the temperature of 450°C, quasi-spherical and homogeneous particles are obtained. The structures elaborated at 450°C show an average size of 422 nm (**Figure 7(b)**). This amelioration is attributed to the indium oxide absence as confirmed by XRD (**Figure 8**).

6. Indium-catalyzed SiNWs grown by VLS mode

6.1 Influence of indium oxide

To study the indium oxide influence on the grown silicon nanowires, a comparative study is carried on. Silicon nanowires are grown using indium nanoparticles elaborated by the two different annealing processes: the conventional and the RTA annealing. **Figure 9(a)** and **(b)** shows the SEM images of the obtained nanowires. The elaborated indium particles have been treated by hydrogen gas with a flow rate of 60 sccm for 10 min before the silane introduction during 15 min.

As observed, the indium particles elaborated by conventional annealing morphology are enhanced and quasi-spherical particles are obtained. Despite this improvement, the SiNWs' density is very low due to the persistence of indium oxide after the hydrogen treatment as explained in Section 3. The indium oxide forming a shell around indium is decreasing its catalytic effect. To confirm the persistence of indium oxide, XRD of the obtained SiNWs is performed (**Figure 10**). In addition to silicon and indium peaks, we notice the presence of In_2O_3 peaks explaining the obtained low density.

Using indium particles elaborated at 450°C by the RTA process, the SiNWs' density is ameliorated, which is attributed to indium oxide's absence. The quasi-totality of indium particles is active to catalyze the SiNWs' growth.

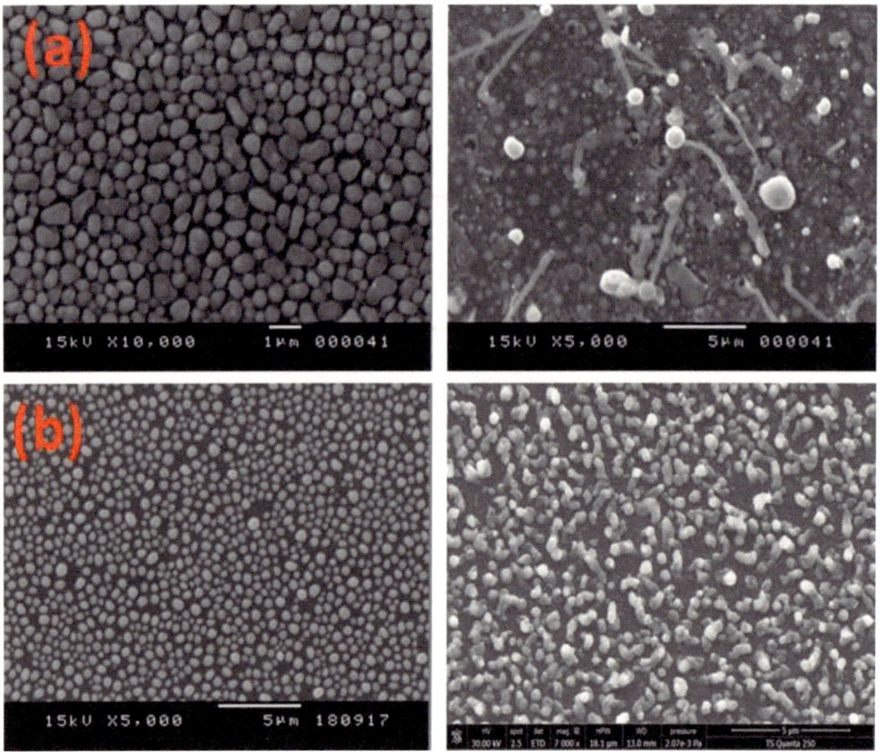

Figure 9.
SEM images of (a) indium particles obtained by conventional annealing and the obtained SiNWs and (b) indium particles obtained by RTA annealing and the obtained SiNWs.

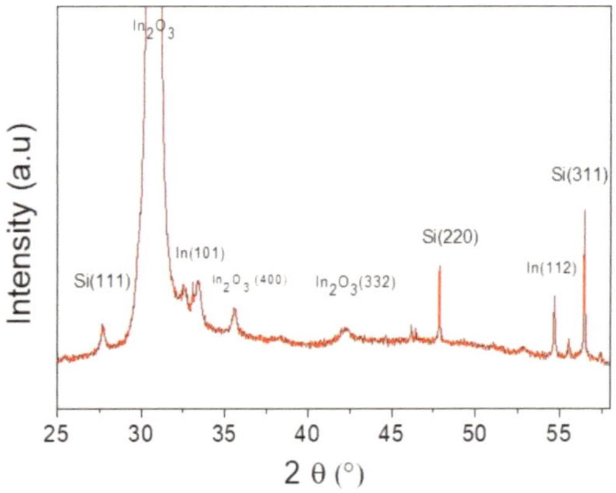

Figure 10.
XRD spectrum of SiNWs catalyzed by indium particles elaborated by conventional annealing.

Indium particles catalyzing the growth are located on the top of the silicon nanowires confirming the VLS mode. Moreover, we notice that the SiNWs' morphology does not seem to depend on the annealing process. The obtained SiNWs in both cases are bent and kinked.

6.2 Indium diffusion in SiNWs

In order to study the indium diffusion in the obtained SiNWs, energy dispersive spectroscopy (EDS) is performed on the top and in the middle of a silicon nanowire. **Figure 11** shows that the wire is consisting of silicon only indicating that indium does not diffuse in the contrary of other catalysts like gold. In addition to that, EDS performed on the catalyst particle shows that the particle is consisting of a indium-silicon alloy as explained by the VLS mode (**Table 1**).

6.3 Indium catalyst removal

Residual indium could be removed by a simple chemical method consisting of soaking samples in 5% hydrochloric acid (HCl) solution (**Figure 12**). The reaction

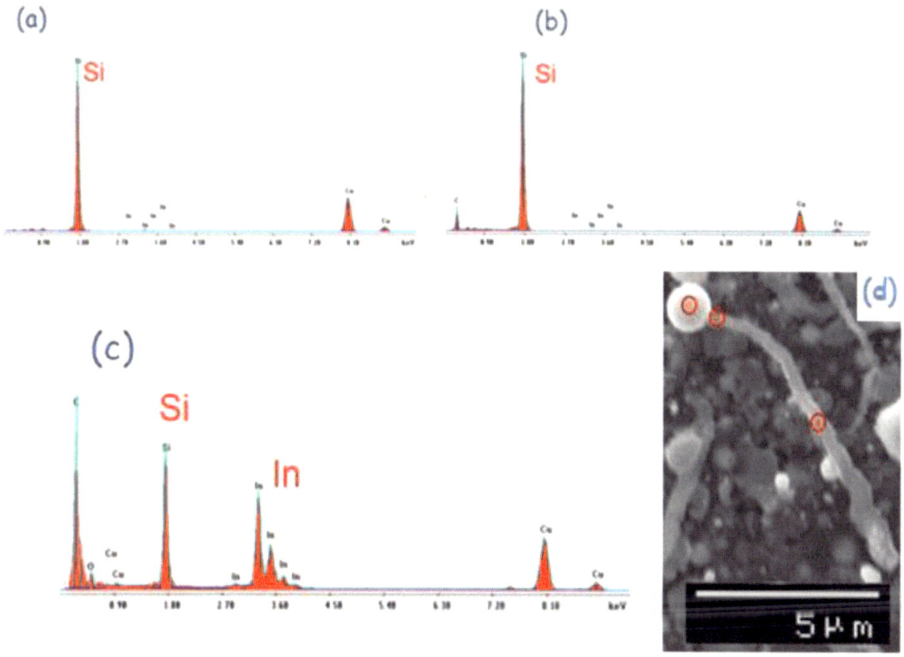

Figure 11.
EDS spectra of a wire performed (a) in the middle, (b) on the top, and (c) on the particle catalyst as shown in (d).

Element (atomic %)	Particle	Top of SiNW	Middle of SiNW
Si	51.1	98.8	99.4
In	48.9	1.2	0.6

Table 1.
Atomic percentage of silicon and indium in a grown silicon wire.

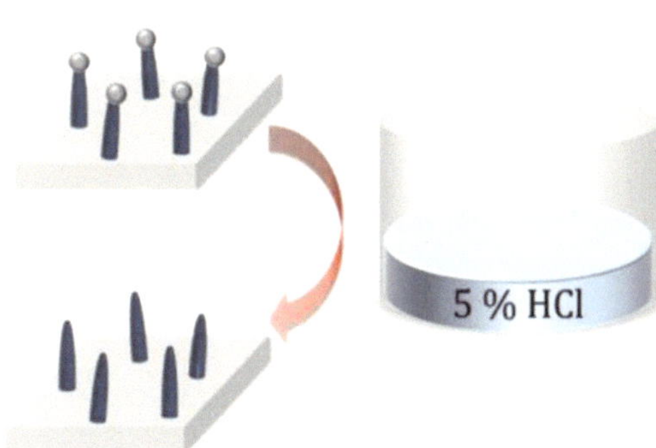

Figure 12.
Illustration of the experimental protocol used.

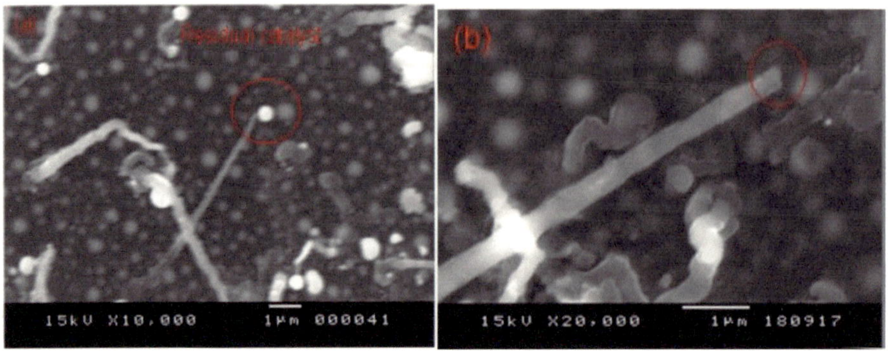

Figure 13.
SEM images of SiNWs (a) before and (b) after catalyst elimination.

between In and HCl could potentially produce two types of indium chloride: $InCl_2$ or $InCl_4$. **Figure 13** shows the grown silicon nanowires before and after indium catalyst elimination.

7. Conclusions

In this work, indium is used as catalyst for silicon nanowires' growth by VLS mode using PECVD. Indium is considered as a promising metal to replace gold as it forms a low-eutectic alloy with silicon and it introduces shallow defects.

Indium catalyst is elaborated by annealing indium-coated silicon substrates using two different annealing processes: conventional and RTA annealing. The annealing conditions influence the catalyst morphology and as a consequence the grown SiNWs. In this work, we have shown that the indium oxide presence is affecting the growth, in particular the density.

Acknowledgements

The author wants to thank Prof. Wissem Dimassi from the Laboratory of Nanomaterials and Systems for Renewable Energy (LaNSER-CRTEn-Tunisia), Prof. Kamel Khirouni from the Laboratory of Physics of Materials and Nanomaterials Applied to Environment (LaPhyMNE—University of Gabès-Tunisia), and Dr. Maroua Yaacoubi Tabassi from the Photovoltaic Laboratory (LPV-CRTEn-Tunisia) for their contributions in this work.

Author details

Rabia Benabderrahmane Zaghouani
Photovoltaic Laboratory, The Research and Technology Centre of Energy, Tunis, Tunisia

*Address all correspondence to: rabia.benabderrahmane@gmail.com

References

[1] Azaiez A, Benabderrahmane Zaghouani R, Khamlich S, Meddeb H, Dimassi W. Enhancement of porous silicon photoluminescence property by lithium chloride treatment. Applied Surface Science. 2018;**441**:272-276. DOI: 10.1016/j.apsusc.2018.02.006

[2] Manai L, Dridi Rezgui B, Benabderrahmane Zaghouani R, Barakel D, Torchio P, Palais O, et al. Tuning of light trapping and surface plasmon resonance in silver nanoparticles/c-Si structures for solar cells. Plasmonics. 2016;**5**:1273-1277. DOI: 10.1007/s11468-015-0171-4

[3] Benabderrahmane Zaghouani R, Manai L, Dridi Rezgui B, Bessais B. Study of silver nanoparticles electroless growth and their impact on silicon properties. Chemistry Journal. 2015;**1**:90-94

[4] Wang J, Liu G, Zhu Q. Indium microrod tags for electrochemical detection of DNA hybridization. Analytical Chemistry. 2003;**75**: 6218-6222. DOI: 10.1021/ac034730b

[5] Aouida S, Benabderrahmane Zaghouani R, Bachtouli N, Bessais B. Hydrogen passivation of silicon nanowire structures. Applied Surface Science. 2016;**370**:49-52. DOI: 10.1016/j.apsusc.2016.02.116

[6] Benabderrahmane Zaghouani R, Aouida S, Bachtouli N, Bessais B. Silver nanoparticles effect on silicon nanowires properties. Chemistry Journal. 2015;**1**:10-14

[7] Tabassi MY, Benabderrahmane Zaghouani R, Khelil M, Khirouni K, Dimassi W. Study of indium catalyst thickness effect on PECVD-grown silicon nanowires properties. Journal of Materials Science: Materials in Electronics. 2017;**13**:9717-9723. DOI: 10.1007/s10854-017-6722-z

[8] Morin C, Kohen D, Tileli V, Faucherand P, Levis M, Brioude A, et al. Patterned growth of high aspect ratio silicon wire arrays at moderate temperature. Journal of Crystal Growth. 2011;**321**:151-156. DOI: 10.1016/j.jcrysgro.2011.02.024

[9] Kamins TI, Williams RS, Basile DP, Hesjedal T, Harris JS. Ti-catalyzed Si nanowires by chemical vapor deposition: Microscopy and growth mechanisms. Journal of Applied Physics. 2001;**89**:1008-1016. DOI: 10.1063/1.1335640

[10] Ke Y, Weng X, Redwing JM, Eichfeld CM, Swisher TR, Mohney SE, et al. Fabrication and electrical properties of Si nanowires synthesized by Al catalyzed vapor-liquid-solid growth. Nano Letters. 2009;**9**: 4494-4499. DOI: 10.1021/nl902808r

[11] Arbiol J, Morral AF, Estrade S, Peiro F, Kalache B, Cabarrocas PR, et al. Influence of the (111) twinning on the formation of diamond cubic/diamond hexagonal heterostructures in Cu-catalyzed Si nanowires. Journal of Applied Physics. 2008;**104**:064312-064317. DOI: 10.1063/1.2976338

[12] Conesa-Boj S, Zardo I, Estrade S, Wei L, Alet PJ, Cabarrocas PR, et al. Defect formation in Ga-catalyzed silicon nanowires. Crystal Growth & Design. 2010;**10**:1534-1543. DOI: 10.1021/cg900741y

[13] Baron T, Gordon M, Dhalluin F, Ternon C, Ferret P, Gentile P. Si nanowire growth and characterization using a microelectronics compatible catalyst: PtSi. Applied Physics Letters. 2006;**89**:233111. DOI: doi.org/10.1063/1.2402118

[14] Zhang Y, Ago H, Liu J, Yumura M, Uchida K, Ohshima S, et al. The synthesis of In, In_2O_3 nanowires

and In_2O_3 nanoparticles with shape-controlled. Journal of Crystal Growth. 2004;**264**:363-368. DOI: 10.1016/j.jcrysgro.2004.01.025

[15] Rodriguez-Sanchez I, Blanco MC, Lopez-Quintela MA. Electrochemical synthesis of silver nanoparticles. The Journal of Physical Chemistry B. 2000;**104**:9683-9688. DOI: 10.1021/jp001761r

[16] Chou NH, Ke XL, Schiffer P, Schaak RE. Room-temperature chemical synthesis of shape-controlled indium nanoparticles. Journal of American Chemical Society. 2008;**130**:8140-8141. DOI: 10.1021/ja801949c

[17] Ganeev RA, Ryasnyanskiy AI, Chakravarty U, Naik PA, Srivastava H, Tiwari MK, et al. Structural, optical and nonlinear optical properties of indium nanoparticles prepared by laser ablation. Applied Phyics B. 2007;**86**:337-341. DOI: 10.1007/s00340-006-2526-1

[18] Xie X, Zeng X, Yang P, Wang C, Wang Q. In-situ formation of indium catalysts to synthesize crystalline silicon nanowires on flexible stainless steel substrates by PECVD. Journal of Crystal Growth. 2012;**347**:7-10. DOI: 10.1016/j.jcrysgro.2012.03.011

[19] Chong SK, Goh B, Dee C, Rahman S. Effect of substrate to filament distance on formation and photoluminescence properties of indium catalyzed silicon nanowires using hot-wire chemical vapor deposition. Thin Solid Films. 2013;**529**:153-158. DOI: 10.1016/j.tsf.2012.07.098

[20] Convertino A, Cuscun M, Nicotra G, Spinella C, Felisari L, Fortunato G, et al. Low-temperature growth of In-assisted silicon. Journal of Crystal Growth. 2011;**335**:10-16. DOI: 10.1016/j.jcrysgro.2011.09.009

[21] Iacopi F, Vereecken PM, Schaekers M, Caymax M, Moelans N, Blanpain B, et al. Plasma-enhanced chemical vapor deposition growth of Si nanowires with low melting point metal catalysts: An effective alternative to Au-mediated growth. Nanotechnology. 2007;**18**:505307. DOI: 10.1088/0957-4484/18/50/505307

[22] Kumar RR, Rao KN, Phani AR. Growth of silicon nanowires by electron beam evaporation using indium catalyst. Materials Letters. 2012;**66**:110-112. DOI: 10.1016/j.matlet.2011.08.064

[23] Legallais M, Nguyen TTT, Mouis M, Salem B, Robin E, Chenevier P, et al. An innovative large scale integration of silicon nanowire-based field effect transistors. Solid-State Electronics. 2018;**143**:97-102. DOI: 10.1016/j.sse.2017.11.008

[24] Le Borgne B, Pichon L, Salaun AC, Le Bihan B, Jolivet-Gougeon A, Martin S, et al. Bacteria electrical detection using 3D silicon nanowires based resistor. Sensors and Actuators B: Chemical. 2018;**273**:1794-1799. DOI: 10.1016/j.snb.2018.07.101

[25] Dytrych P, Drinek V, Bumba J, Kastanek F, Solcova O. Silicon nanowires based photoanode for hydrogen evolution. International Journal of Hydrogen Energy. 2018;**43**:18136-18141. DOI: 10.1016/j.ijhydene.2018.04.244

[26] Massalski TB, Okamoto H, Subramanian PR, Kacprzak L. Binary Alloy Phase Diagrams. 2nd ed. Ohio: ASM International, Materials Park; 1990

[27] Benabderrahmane Zaghouani R, Tabassi MY, Khirouni K, Dimassi W. Vapor–liquid–solid silicon nanowires growth catalyzed by indium: Study of indium oxide effect. Journal of Materials Science: Materials in Electronics. 2019;**30**:9758-9766. DOI: 10.1007/s10854-019-01311-2

Chapter 5

Elastic, Optical, Transport, and Structural Properties of GaAs

Thamer A. Tabbakh, Raghad M. Aljohany, Hatem Alhazmi and Rawan M. Alsulami

Abstract

One of the major objectives of physics is to understand the physical properties of compound metals. Based on this very objective, in this chapter, we intend to review the physical as well as chemical properties of Gallium Arsenide material.

Keywords: GaAs, basic principle, physical properties, optical properties, chemical properties, physics

1. Introduction

Each time the creation and refinement of new material growth techniques give rise to the finding of new equipment. The semiconductor which has been found to be most interesting of all (when used), relates to the applications in electronic devices for instance, high speed circuits formed of group III-V compounds along with switches or amplifiers. One of the types of is the Gallium Arsenide (GaAs) [1, 2]. A III-V semiconductor that it is, its composition includes gallium which is an element of the group 3 having orthorhombic trigonal crystal structure coupled with arsenic which belongs to the group 5 and has a trigonal crystal structure. This very combination leads to interesting physical as well as chemical properties in this semiconductor [3, 4]. With different electronic band gaps, these semiconductors can, at ambient temperatures, crystalize into a zinc-blende cubic type crystal structure. Post the discovery of transistor in the year 1947, Gallium arsenide (GaAs) has presented itself as a prominent material for electronic devices. The technological importance and the need of study of Gallium arsenide (GaAs) in the last few years are due to its high melting point at 1238° C along with a density of 5.3176 g/cm3. It is obtained as a by-product from the extraction of ores of zinc and aluminum and is present at 5-15 mg/Kg in Earth's crust [5, 6].

GaAs is the basis of a worldwide industry it has a wide usage area changing from microwave frequency integrated circuits to optical windows. Among III-V semiconductors, GaAs has remained useful as a semiconductor material, widely being used in optoelectronic and microelectronic devices. Given this wide use of GaAs, the idea came to write this chapter focusing on the physical and chemical properties of this important material. While excluding the impurities present and their associated effects, the focus is on the characteristics of GaAs [7, 8].

Included in this chapter, are a few physical as well as chemical properties of gallium arsenide (GaAs) such as the crystal Structure, the direct bandgap of GaAs with its zinc blende type crystal structure, electronic structure, light-emitting properties, electromagnetic properties, photovoltaic properties, phase transition and mechanical as well as elastic properties. The geometric structures, stabilities, and electronic properties like higher saturated electron velocity and high carrier mobility with a small dielectric constant and high resistivity. This chapter also elaborates on the thermodynamic properties of GaAs such as thermal expansion and thermal conduction.

These great properties led to the production of new and unique devices like high-efficiency light emitters, light sensors, and high-speed switching devices. The GaAs is considered as an outstanding member of the III-V semiconductor family. It has many exceptional features, especially for the recent optoelectronic industry. Hence, special focus has been laid on the examination of physical properties of this material [9–11].

2. GaAs and semiconductors direct bandgap concept

The band structure is the major part of the semiconductors. Briefly explaining it, at absolute zero, a bandgap or an energy gap separates the conduction band (lowest empty band) with the valence band (highest filled band). Therefore, at T = 0, electricity is not conducted by the material. The electrons are enabled to be excited into the conduction band through several processes, such as optical absorption or thermal excitation, at finite temperatures and electrical conduction is allowed as there are empty states in the valence band. Energy, in the forms of heat or photons, is released when the electrons return to the valence band [12, 13].

As mentioned in **Figure 1**, two types of bandgaps are there based on different conditions. The first one is if, over the top of the valence bond, the bottom of the conduction band does not rest. As a result of this, it is called an indirect gap. Also, photon is necessary in order to provide the momentum required to reach the state in the conduction band, and the electronic transition to happen. However, in case of GaAs and other direct band type semiconductor, the bottom of the conduction band site and the top of the valence band are on top of each other, Therefore even without a change in the wave vector, the electron is able to get excited from the

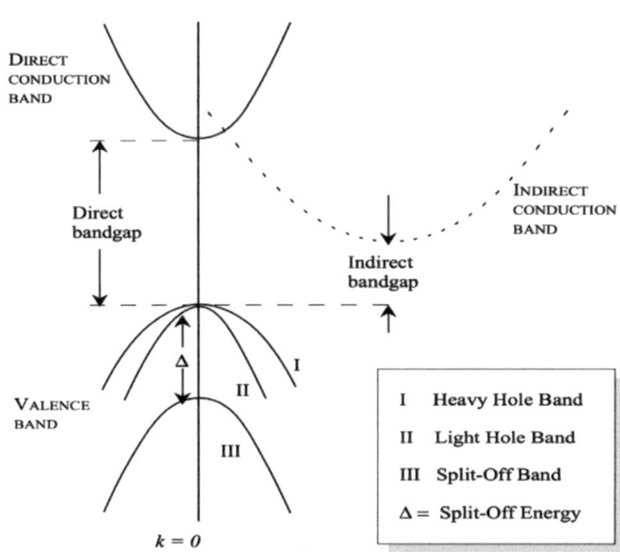

Figure 1.
Schematic of the valence band, direct bandgap, and indirect bandgap conduction bands [13].

valence band. A photon on absorbing required energy is sufficient enough for this. Moreover, through the emission of a photon, transition to the valence band from the conduction band can easily be done by the electron. While no interaction of photon is required, emission of light energy of the desired wavelength of 850 nm bandgap occurs and it allows the direct band recombination of holes and electrons. In the absence of defects, the energy released by the dominant mechanism of the indirect bandgap is by photons via electromagnetic radiation. However, photons release energy in the form of heat in the case of indirect bandgap semiconductors [14–16].

Being a good optoelectronic material, direct bandgap in GaAs is considered a useful material in the field of optoelectronics and other electronic fields and is used comprehensively in semiconductor lasers as well as light-emitting diodes. Its use has also been encouraged in the making of high-efficiency solar cells, Gunn diodes, Infrared LEDs, solid-state detectors and radar systems. We have a classification of 1–3 categories depending upon the magnitude of bandgap energy of the materials, namely, narrow, mid and wide-bandgap. Of all the properties of a semiconductor, the presence of energy gap leads all others [16, 17].

Not only band-gap engineering permits the making of band diagrams that have continuous as well as arbitrary band-gap variations, but it is also considered among the strongest tools for the new semiconductor devices and materials. For a specific application, the transport properties of holes, as well as electrons, may be continuous and independent. This approach leads us to a new generation of devices having unique capabilities ranging from resonant tunneling transistors to solid-state photomultipliers. More than for any semiconductor, many band structures for GaAs are precisely known. **Figures 2** and **3** make it clear, showing 1.519 eV as the fundamental energy gap for Gallium Arsenide. Also, the high-temperature performance of GaAs is largely attributed to its wide bandgap [20–22].

Other than this, photoconductivity, a feature semiconductor exhibit under suitable trial conditions is another benefit associated with the bandgap [23]. This occurs when an increase in electrical conductivity happens when an incident light falls on a semiconductor. The suitability of the semiconductor material in optoelectronic devices is also decided by its photoconductive response. Such materials which respond well to the photoconductivity, find themselves useful in the making of infrared sold state detectors. Also, in GaAs, the transport properties of hot electrons

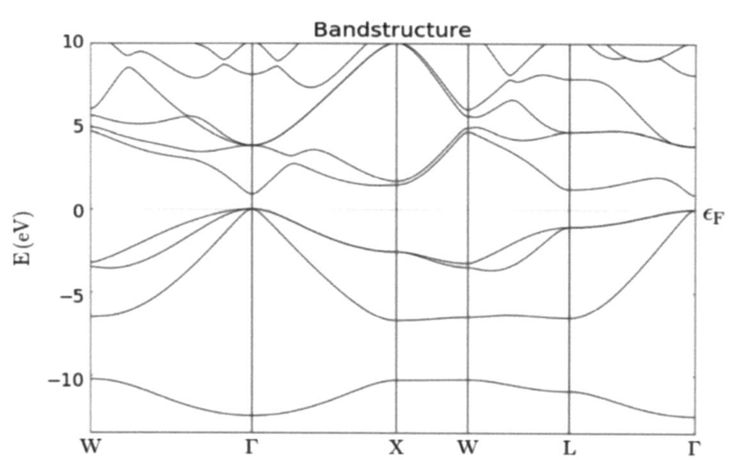

Figure 2.
Structure of GaAs energy band gap [18].

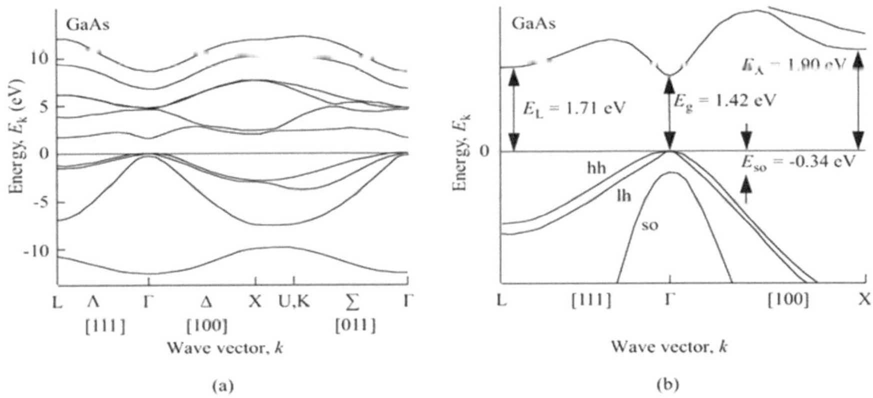

Figure 3.
(a) Calculated band structure of GaAs using the tight binding method. (b) Sketch of GaAs band structure near the Γ-symmetry point showing conduction band, heavy hole band (hh), light hole band (lh), and split-off band (so) in the X [100] and L [19].

are largely affected due to the bandgap. Alloying is another controllable bandgap which useful property of the GaAs [22, 24–26].

The usefulness of GaAs and its alloys in optoelectronics such as sold state lasers and LEDs lies in the bandgap of GaAs which, in the infra-red range results in the emission of photon. The ability of GaAs to retain their semiconductor properties at high temperatures, giving stability to the GaAs comes from the wider bandgap of GaAs [19]. **Figure 3** shows that Calculated band structure of GaAs using the tight binding method with different semiconductor materials.

3. GaAs and semiconductor carrier density

That GaAs is an extremely poor conductor is corroborated by the fact that GaAs has low flow intrinsic carrier density when present in an undoped or pure form. Hence it is mostly considered as semi-insulating. Adding the dopants of either the p- or the n- that is the positive and the negative types respectively, alters this property. Many active devices have been able to be made on a single substrate due to this semi-insulating property, where each device's electrical isolation is provided by the GaAs. For contraction of the electronic circuitry, this characteristic has been found be quite important [27, 28].

With regards to the transport properties, two important questions that come up where the incorporation of dopants has been done – a) mobility, b) effective carrier concentration [29, 30].

$$n_i = \sqrt{N_C N_V}\ exp\left(-\frac{E_g}{K_B T}\right) \quad (1)$$

$N_c$and N_V, the effective density of states at the band edges, are dependent on temperature and the effective mass of the electron and holes respectively.

$$N_C = 2\left(\frac{2\pi m^*{}_e K_B T}{h^2}\right)^{\frac{3}{2}} \quad (2)$$

$$N_v = 2\left(\frac{2\pi m^*{}_h K_B T}{h^2}\right)^{\frac{3}{2}} \quad (3)$$

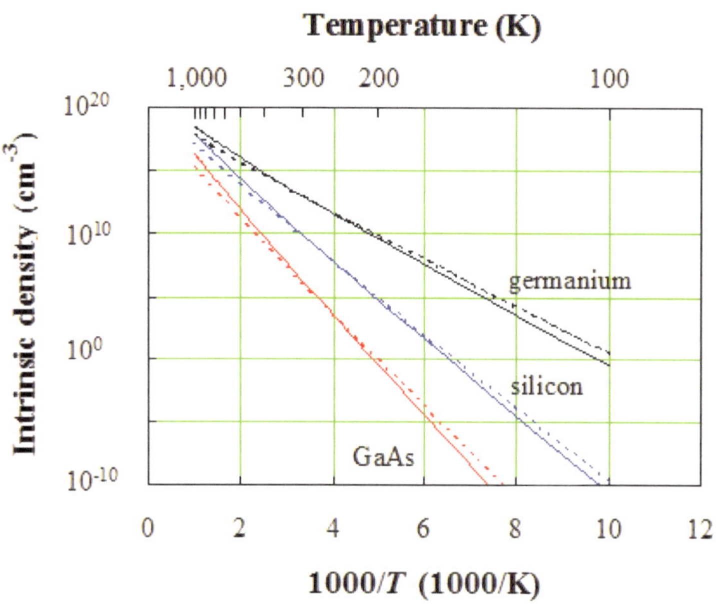

Figure 4.
Bandgap energy inverse T for Ge, Si, GaAs in the range 200–1000 K. T [31].

The conductivity is given by

$$\sigma_i = n_i e(\mu_e + \mu_h) \tag{4}$$

Eq. (1) and **Figure 4** can be used to understand the effect of temperature on the concentration of the carrier. It shows the entry of temperature in both exponential as well as pre-exponential terms. There are two effects of increased temperature [31, 32]:

- the effective density of states of the edge of the band is increased (N_V and N_C).
- the exponential portion of Eq. 10 shows a decrease, since its denominator is T.

4. GaAs crystal structure

In solid-state physics, the central theoretical problem happens to be the determination of the energy bands. In other words, in case of solids, the central theoretical problem is the calculation of energy levels of electrons. In order to calculate physical properties such as mechanical properties, magnetic order, optical dielectric or the vibrational spectra, in principle knowledge of the electrons and the energies associated with them is required. In contrast, calculation of lattice constants and other bulk ground state properties such as atomic positions and bulk modulus is considered to be important in physics associated with condensed matter. Such bulk calculations not only help understand as well as characterize the mechanical properties of the matter, but they also help predict their properties in extreme conditions [33–35].

Having a basis and a cubic lattice that is two face centered, the structure of crystal of GaAs is a zincblende structure or cubic sphalerite. In the classic basis, at the origin of the lattice, there is one GaAs molecule. From (0,0,0), the vector of one atom to another at (1/4,1/4,1/4) of the molecule constitutes the basis. Two FCC lattices, one of as while the other being of Ga can also form a crystal as shown in

Figure 5. While arsenic atoms are represented in orange color, the Gallium ones are shown in purple color. As shown, there are 4 arsenic atoms against 14 gallium atoms. This makes a tetrahedral bond, similar to the one in a diamond lattice, but replaced with Ga and as where each Ga is connected to four other atoms. It shows ionic bonding with the presence of two types of atoms [37–39] (**Figure 6**).

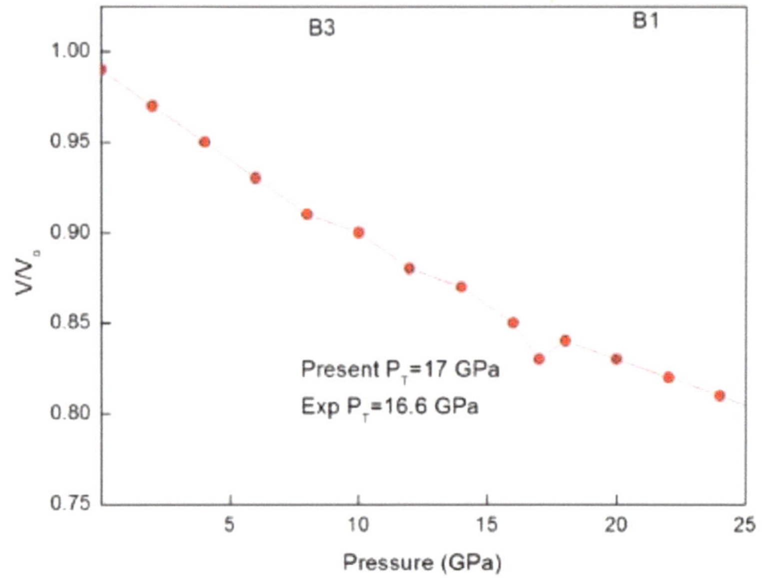

Figure 5.
P-V plot of GaAs up to 25 GPA [36].

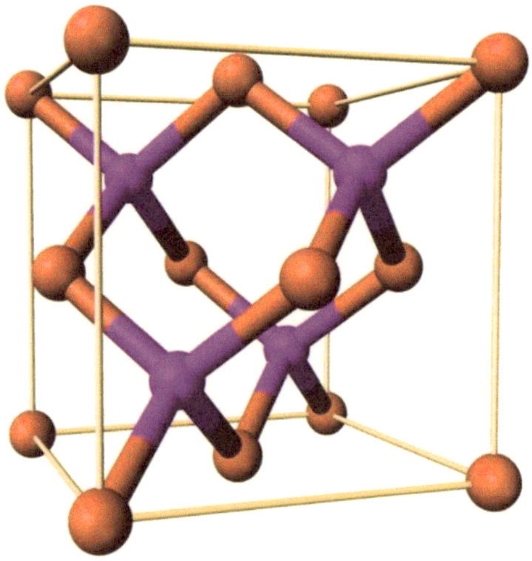

Figure 6.
The crystal structure of GaAs cubic unit cell of GaAs [38].

5. GaAs Fermi level and lattice constant

Figure 7 shows that the drawing of the energy band of the semiconductor, which demonstrates the bandgap along with valence and conduction electrons. An electron leaves a hole in the valence band on moving to the conduction band from

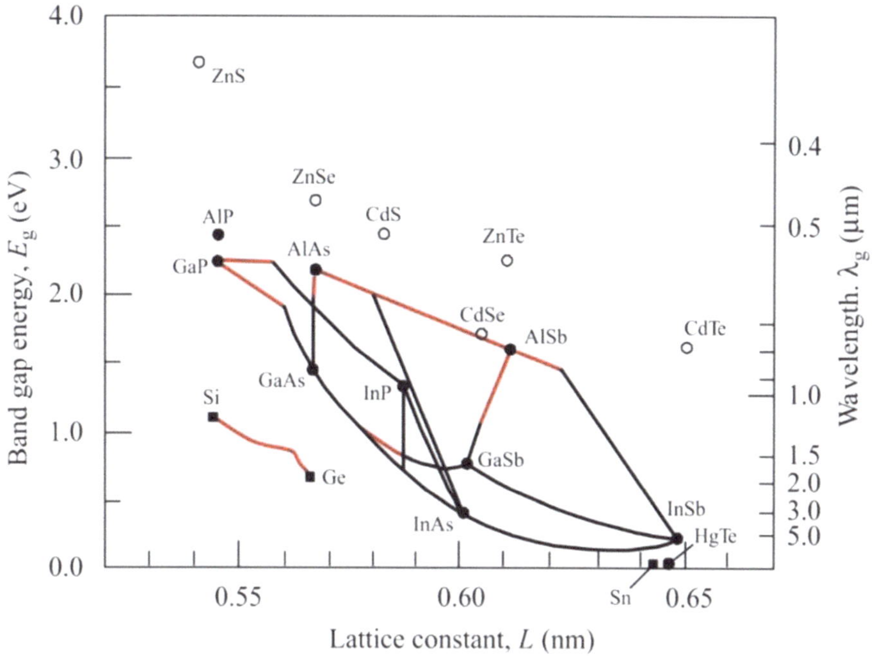

Figure 7.
Room-temperature band gap energy, Eg, as a function of lattice constant for several semiconductors. Lines connecting binary compounds such as GaAs and AlAs represent alloy composition with either a direct band gap (thick solid line) or indirect band gap (thin red line). The III–V and II–VI semiconductor compounds in the figure have the zinc blende crystal structure. Si and Ge have the diamond crystal structure [40].

the valence band. For the other electrons in the valence band energy levels, this hole is an empty state and behaves in a manner similar to a $+v_e$ charged particle in the valence band. The number of electrons in the conduction band per unit volume, written as cm^3, is counted to quantify the electron concentration, and is represented by 'n' [41, 42].

The density of holes in the valence band equals the density of electrons in the conduction band.

$$n = 1$$

$$N_C = \exp\left[\frac{E_F - E_C}{K_B T}\right] = N_v\, exp\left[\frac{E_v - E_F}{K_b T}\right] \tag{5}$$

In the above formula,
E_F - Fermi energy.
N_V - effective density of states in the valence band.
E_V - valence band edge.
k_B - Boltzmann's constant.
E_C - conduction band edge.
T - temperature in K.
N_C - effective density of states in the conduction band.
Reordering the above equation, we have

$$exp\left(\frac{2E_F - E_C - E_V}{K_B T}\right) = \frac{N_V}{N_C} \tag{6}$$

Taking log we have

$$\left(\frac{2E_F - E_C - E_V}{K_B T}\right) = In\frac{N_V}{N_C} \tag{7}$$

Solving for E_F

$$E_F = \frac{E_C + E_V}{2} + \frac{K_B T}{2} In\left(\frac{N_V}{N_C}\right) \tag{8}$$

It is in the middle of the bandgap that we have the Fermi energy $(E_C + E_V)/2$. Therefore, as shown in 7, the energy bandgap latticed matched for several semiconductors that is related to GaAs and band alignments for sever III-V semiconductor with GaAs (**Figure 8**).

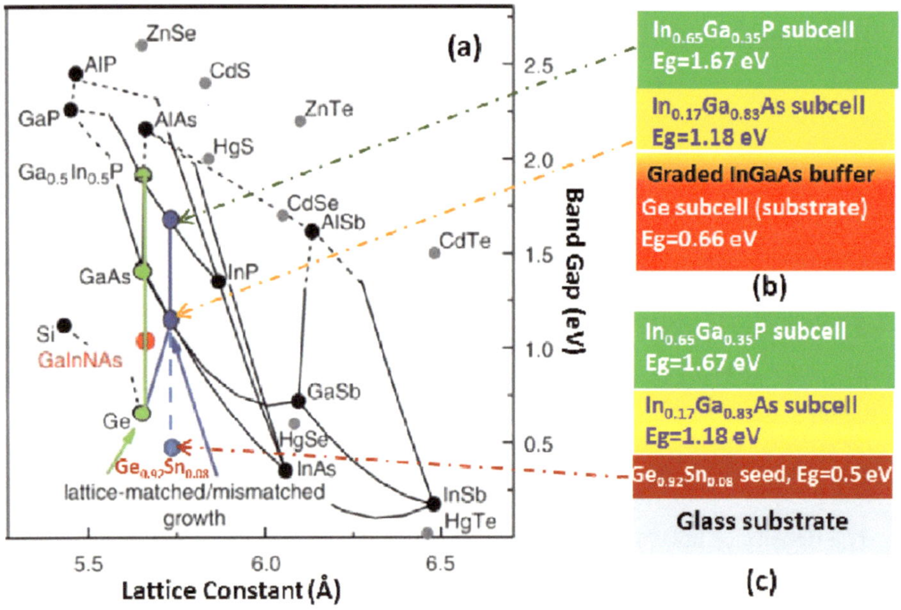

Figure 8.
Band alignments for sever III-V semiconductor with GaAs. (a) Lattice constants and band gaps of different semiconductor materials. The lattice constants of Ge and GaAs are close to a variety of semiconductors with different band gaps, thereby commonly used as substrates for tandem cells. (b) Schematics of a triple-junction tandem cell with optimized compositions of InGaAs and InGaP, which correspond to the blue dots in (a). A buffer layer has to be applied to accommodate the lattice mismatch between the Ge substrate and In 0.17 Ga 0.83 As; (c) Proposed latticematched GeSn/InGaAs/InGaP tandem cell [42].

6. GaAs transport properties

An ionic bond exists between the electrons in the valence band of GaAs atoms. Therefore, in solids, they are not free to transport. However, they can move through a solid if the electron gets excited to the conduction band. Hence, an electron leaves a hole in the valence band on moving to the conduction band from the valence band. An electron jump from one bond to another enables the hole to move in the valence band. Further, these holes and electrons can move upon affected by an electric field [43–45].

The equation that helps derive the acceleration of the electrons (ae) is-

$$a_e = \frac{eE_x}{m_e} \tag{9}$$

Where $m_e$is the electron's rest mass and E_X is the electric field. Under an electric field, the interaction of electrons with the solid atoms should also be considered.

In the case of an electron in a solid, under an external field, the interaction with the solid atoms should also be taken into account. Let this interaction of electrons with the atoms of the solid be summed up as $\sum f_{int}$ [46, 47].

$$a_e = \frac{eE_x + \sum f_{int}}{m_e} \tag{10}$$

The conduction electrons are initially in the lowest energy valley, Γ minimum. Here, they are distinguished by a low effective mass and have high mobility. On applying the electric field, these electrons are rapidly accelerated by the field to high velocities: they gain kinetic energy. The electrons' ability gains enough energy to transfer from this valley to the next higher valley, the L minimum. These upper valleys are distinguished by a larger effective mass, and consequently both lower electron mobility and a greater density of available electron positions. The great density of states encourages transfer to these valleys of electrons with suitable energy. There is a fall in the average velocity of all the electrons as the energetic electrons transfer to the upper levels. There is much work available which is focused on the calculation of hot electron transport properties in GaAs. Gallium arsenide exhibits the `transferred electron' which is commonly known as the TE effect. This transfer of electrons from one region to another energy band structure is an electric-induced field transfer. It has high electron mobility, with a negative resistance being observed and has a small dielectric constant. This is primarily because of extensive utilization of GaAs in ultrahigh frequency, high-temperature resistance and low power circuits and devices [18–53].

$$average\ velocity\ \nu = \frac{n_\Gamma \nu_\Gamma + n_r \nu_r}{n_\Gamma + n_l} \tag{11}$$

7. GaAs optical properties

There are two main categories in which the optoelectronic devices can be categorized. First is the set of those devices in which electrical current get converted into electromagnetic radiation i.e. light. Second is the set of those devices in which light is converted into electric current. An example of the first category of devices is the LEDs. Optoelectronic devices as well as light-emitting devices have GaAs in them. Macroscopic evidence is available for the interaction of matter with light and it is made of four components, namely, incident, reflected, transmitted, and scattered component [52, 54, 55].

Absorption of a photon in a semiconductor can happen in a number of ways. This is commonly known as interbond absorption. This happens when in the conduction band an electron is excited up after having absorbed a photon in the valence band. The momentum is unchanged in case of direct gap semiconductor while there is an increase in electron's energy. The momentum is though shifted in case of an indirect gap material. This is made possible by a decrease in bandgap energy with the increase in temperature and a photon being either absorbed or emitted. The two factors responsible for this are lattice vibration (phonons) and thermal expansion. The increase of lattice constant is due to the thermal expansion. This further leads to

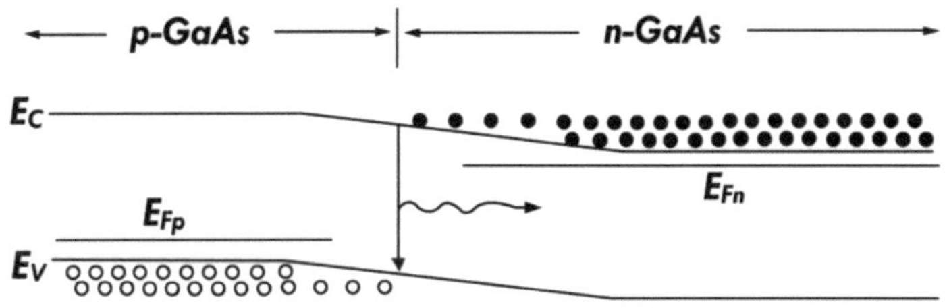

Figure 9.
The heterojunction has a higher quantum efficiency since the carriers are localized GaAs [63].

the change in the periodic potential as seen by the electron. The band structure is altered due to these changes [56–59].

Temperature and bandgap shift are related as per the following formula

$$E_g = E_g - \frac{\propto T^2}{T + \beta} \tag{12}$$

where
Eg - band gap energy
T - temperature
Eg (0) - bandgap at 0 K having units same as that of energy
α has units the same as that of temperature/energy.

Prominently used in optoelectronic as well as microelectronic devices, Gallium arsenide, GaAs is considered a good semiconducting material, having high electron mobility. In the semiconductor material, through the interaction of electron and photons, sunlight is converted into electricity directly by photovoltaic cells [60–62].

Electronic excitation of luminescence is the reason, it is also called as optical radiation. State energy is emitted in the form of EM radiation when the excited electrons move back to the ground. Depending upon the electronic excitation created originally, there are four different types of luminescence, namely,

$$\text{Photoluminescence} - \text{incident light} \rightarrow \text{electronic excitation}$$

$$\text{Radioluminescence} - \text{ionizing radiation } (\beta - \text{rays}) \rightarrow \text{electronic excitation}$$

$$\text{Cathodoluminescence} - \text{electron beam} \rightarrow \text{electronic excitation}$$

$$\text{Electroluminescence} - \text{electrical field} \rightarrow \text{electronic excitation}$$

The functioning of LEDs is by electroluminescence. Using a functional bias, electric current which includes holes and electrons are forwarded to the device. Light is emitted by the recombination of these holes and electrons. In order to increase efficiency, heterostructure LEDs are used. The holes and the electrons, collectively called the carriers are confined in a small spatial region in order to achieve this. Due to the localization of carriers in GaAs, there is higher quantum efficiency at the heterojunction. Hence, it is only in the i-GaAs region, that the recombination takes place. **Figure 9** draws a clear picture of the emitted wavelength in case of both LEDs as shown in **Figure 9** [63–66].

8. GaAs elastic properties

The physical knowledge of the materials namely the phase transitions, interatomic forces and the mechanical features apart from many other features are better understood by examining the elastic properties of the material.

8.1 Density

Being low-density materials, III-V compounds, under pressure, have the ability to show transitions of structural phase to a phase with higher density. **Figure 3** clearly shows that with increase in pressure, the density of GaAs also increases [67].

8.2 Phase transition

Figure 5 shows the P-V graph which helps determine the B3 → B1 phase transition of GaAs. At 17 GPa there is an unexpected decrease in the GaAs's volume. This decrease is attributed to the change in the structural phase related with phase transition of B3 → B1 [23, 36, 68–71].

9. Conclusion

Gallium Arsenide is deemed as an eminent member of the III-V semiconductor group. It has presented itself as a notable material for electronic devices and vastly being used in optoelectronic and microelectronic devices. The precept knowledge of the electrons and the energies related to GaAs is required in order to calculate physical properties. Furthermore, calculation of lattice constants and other bulk ground state properties is considered to be paramount in physics associated with condensed matter.

In this chapter we have reported the basic principle for the GaAs material as well as the physical, optical, and chemical properties. Also, we have reported the crystal growth of the GaAs with another semiconductor material such as In and Al.

Author details

Thamer A. Tabbakh[1], Raghad M. Aljohany[1]*, Hatem Alhazmi[2] and Rawan M. Alsulami[1]

1 Material Science Institute, King Abdulaziz City for Science and Technology, Riyadh, Saudi Arabia

2 National Center for Environmental Technology, King Abdulaziz City for Science and Technology, Riyadh, Saudi Arabia

*Address all correspondence to: raljohany@kacst.edu.sa

References

[1] Aleshkin VY, Dubinov AA. Direct band Ge and Ge/InGaAs quantum wells in GaAs. J Appl Phys. 2011;109(12):4–9.

[2] Bickford JR. Doped Aluminum Gallium Arsenide (AlGaAs)/ Gallium Arsenide (GaAs) Photoconductive Semiconductor Switch (PCSS) Fabrication. 2016;

[3] Scrap A, Cheng T, Liu C, Tsai T, Shen Y. A Process for the Recovery of Gallium from Gallium. 2019;

[4] Fang SF, Adomi K, Iyer S, Morkoç H, Zabel H, Choi C, et al. Gallium arsenide and other compound semiconductors on silicon. J Appl Phys. 1990;68(7).

[5] Kasap S, Capper. Gallium Arsen. 2006;499–536.

[6] Kurt HH. The Features of GaAs and GaP Semiconductor Cathodes in an Infrared Converter System. 2018;(May 2017).

[7] Kurt HH, Tanrıverdi E. The Features of GaAs and GaP Semiconductor Cathodes in an Infrared Converter System. J Electron Mater. 2017;46(7): 4024–33.

[8] Price GL. Gallium Arsenide and Related Compounds. Met forum. 1984;7 (3):171–86.

[9] Güler E, Güler M. Phase Transition and Elasticity of Gallium Arsenide under Pressure 2 . Computational Details. 2014;17(5):1268–72.

[10] Nugroho MB. Semoconductor Bandstructure. J Chem Inf Model. 2013; 53(9):1689–99.

[11] Ashby, C., Baca, A. and Institution of Electrical Engineers., 2005. Fabrication Of Gaas Devices (EMIS Processing Series ; No. 6). Institution of Engineering and Technology - IET.

[12] Electronic Materials, Devices, and Fabrication (Prof. S. Parasuraman, IIT Madras): Lecture 02 - Introduction to Semiconductors

[13] Aleshkin, V. Y. and Dubinov, A. A. (2011) 'Direct band Ge and Ge/InGaAs quantum wells in GaAs', Journal of Applied Physics, 109(12), pp. 4–9. doi: 10.1063/1.3594753.

[14] Capasso, F. (1987) 'Band-Gap Engineering: From Physics and Materials To New Semiconductor Devices.', Science, 235(4785), pp. 172–176. doi: 10.1126/science.235.4785.172.

[15] Nayak, C. (2004) 'Quantum Condensed Matter Physics', Lecture Notes, pp. 1–17. Available at: http://stationq.cnsi.ucsb.edu/~nayak/QFT-9-29-2011.pdf.

[16] WANG, F. F. Y. (1989) 'Carrier Concentrations', Introduction to Solid State Electronics, pp. 139–153. doi: 10.1016/b978-0-444-87317-0.50013-7.

[17] Copple A, Ralston N, Peng X. Engineering direct-indirect band gap transition in wurtzite GaAs nanowires through size and uniaxial strain

[18] Nugroho MB. Semoconductor Bandstructure. J Chem Inf Model. 2013; 53(9):1689–99

[19] Transport E. Density of States, Fermi Energy, and Energy Bands. Thermoelectr Des Mater. 2016;189–205.

[20] Williamson AJ, Franceschetti A, Fu H, Wang LW, Zunger A. Indirect band gaps in quantum dots made from direct-gap bulk materials. J Electron Mater. 1999;28(5):414–25.

[21] Electronic Materials, Devices, and Fabrication (Prof. S. Parasuraman, IIT Madras): Lecture 02 - Introduction to Semiconductors.

[22] Diakite YI, Traore SD, Malozovsky Y, Khamala B, Franklin L, Bagayoko D. Accurate Electronic, Transport, and Bulk Properties of Gallium Arsenide (GaAs).

[23] Owolabi JA, Onimisi MY, Abdu SG, Olowomofe GO. Determination of Band Structure of Gallium-Arsenide and Aluminium-Arsenide Using Density Functional Theory. Comput Chem. 2016;04(03):73–82.

[24] F. Rana Chapter 1 Review of Basic Semiconductor Physics.

[25] Lou CY, Somorjai GA. Lawrence Berkeley National Laboratory Recent Work Title STUDIES OF THE VAPORIZATION MECHANISM OF GALLIUM ARSENIDE SINGLE CRYSTALS.

[26] August CL. Carrier Concentrations.

[27] Fermi energy of an intrinsic semiconductor [Internet]. Available from: https://lampx.tugraz.at/~hadley/psd/weblectures/Ef_intrinsic/index.php

[28] Ashby, C., Baca, A. and Institution of Electrical Engineers., 2005. Fabrication Of Gaas Devices (EMIS Processing Series ; No. 6). Institution of Engineering and Technology - IET.

[29] Cheng TH, Liu CJ, Tsai TY, Shen YH. A process for the recovery of gallium from gallium arsenide scrap. Processes. 2019;7(12).

[30] Güler E, Güler M. Phase Transition and Elasticity of Gallium Arsenide under Pressure. Computational Details. 2014;17(5):1268–72.

[31] Ghita R V, Negrila CC, Cotirlan C, Logofatu C. ON THE PASSIVATION OF GaAs SURFACE BY SULFIDE COMPOUNDS. Vol. 8, Digest Journal of Nanomaterials and Biostructures.

[32] Kayali S. I. GaAs Material Properties. Kabita K, Maibam J, Sharma BI, Brojen Singh RK, Thapa RK. Density functional theory study on pressure induced structural transformation, elastic properties and electronic structure of gallium arsenide (GaAs) . Vol. 8, International Journal of Innovation and Applied Studies. 2014.

[33] Flannery CM, Chilla E, Semenov S, Froehlich HJ. Elastic properties of GaAs obtained by inversion of laser-generated surface acoustic wave measurements. In: Proceedings of the IEEE Ultrasonics Symposium. IEEE; 1999. p. 501–4

[34] Boulder, Chapter 2: Semiconductor Fundamentals. August 2007

[35] Si F. Lecture 5 : Intrinsic Carrier concentration in intrinsic Si Carrier concentration in intrinsic semicon-. 2 (2):1–5.

[36] Diakite YI, Traore SD, Malozovsky Y, Khamala B, Franklin L, Bagayoko D. Accurate Electronic, Transport, and Bulk Properties of Gallium Arsenide (GaAs).

[37] Iskakova, Kulpash, Rif Akhmaltdinov, and Aisara Amanova. "About the energy levels of GaAs." *Journal of Physics: Conference Series (см. в книгах)*. Vol. 510. No. 1. Institute of Physics and IOP Publishing Limited, 2014.

[38] Spirkoska D, Arbiol J, Gustafsson A, Conesa-Boj S, Glas F, Zardo I, et al. Structural and optical properties of high quality zinc-blende/wurtzite GaAs hetero-nanowires.

[39] Iskakova K, Akhmaltdinov R, Amanova A. About the energy levels of GaAs. J Phys Conf Ser. 2014;510(1).

[40] Ashby, C., Baca, A. and Institution of Electrical Engineers., 2005. Fabrication Of Gaas Devices (EMIS Processing Series ; No. 6). Institution of Engineering and Technology - IET.

[41] Pearsall TP. Indium Gallium Arsenide Phosphide. In: Encyclopedia of

Materials: Science and Technology. Elsevier; 2001. p. 4048–6

[42] Dilli Z. Intrinsic and Extrinsic Semiconductors , Fermi-Dirac Distribution Function , the Fermi level and carrier concentrations Review : Charge Carriers in Semiconductors. Enee. 2009;313:1–8.

[43] Nugroho MB. Semoconductor Bandstructure. J Chem Inf Model. 2013; 53(9):1689–99

[44] Sharmin M, Choudhury S, Akhtar N, Begum T. Optical and Transport Properties of p-Type GaAs. J Bangladesh Acad Sci. 2012 Jun 17;36(1): 97–107.

[45] Požela J, Reklaitis A. Electron transport properties in GaAs at high electric fields. Solid State Electron. 1980 Sep 1;23(9):927–33.

[46] Guen-Bouazza A, Sayah C, Bouazza B, Chabane-Sari NE. Comparison of electron transport properties in submicrometer InAs, InP and GaAs n+−i-n+ diode using ensemble Monte Carlo simulation. J Mater Environ Sci. 2014;5(4):1238–43.

[47] Požela J, Reklaitis A. Electron transport properties in GaAs at high electric fields. Solid State Electron. 1980 Sep 1;23(9):927–33.

[48] Electronic Materials, Devices, and Fabrication (Prof. S. Parasuraman, IIT Madras): Lecture 02 - Introduction to Semiconductors

[49] Price GL. Gallium Arsenide and Related Compounds. Met forum. 1984;7 (3):171–86.

[50] Leds O. Lecture 16 : Light emitting diodes. :1–13.

[51] Nguyen RH. Gallium arsenide. IEEE Potentials. 1999 Dec;17(5):33–5.

[52] Das C, Begum J, Begum T, Choudhury S. EFFECT OF THICKNESS ON THE OPTICAL PROPERTIES OF GaAs THIN FILMS. Vol. 37, Journal of Bangladesh Academy of Sciences. 2013.

[53] Taheri A, Davoodi M, Setayeshi S. Investigation of electron mobility in GaAs-based devices using genetic algorithm. COMPEL - Int J Comput Math Electr Electron Eng. 2012;31(2):604–18.

[54] Anua NN, Ahmed R, Shaari A, Ul Haq B, Binti Mohamad M. DFT investigations of the optical properties of gallium arsenide. In: Advanced Materials Research [Internet]. Trans Tech Publications Ltd; 2014. p. 429–38

[55] Tanasa C, Lee J, Montoya J, Singh B, Tanasa C. " Real " Semiconductor Project GaAs Part 3. 2001;1–16.

[56] Buonanno PR, Astronomico O, Cnr IFAC, Panciatichi V, Vanni PP. AT T I D E L L A « F ONDA Z ION E GI O RG IO RO NCHI ». 2012;(January).

[57] Al-ammar KHB, Hashim FS, Jappor. Structural and electronic properties of gallium arsenide crystal using INDO method. :1–16.

[58] Tanasa C, Lee J, Montoya J, Singh B, Tanasa C. " Real " Semiconductor Project GaAs Part 1. 2001;1–16.

[59] Surfaces S. GaAs Heterostructures and 2D electron gas. Screening. :8–16.

[60] Brozel M. (2006) Gallium Arsenide. In: Kasap S., Capper P. (eds) Springer Handbook of Electronic and Photonic Materials. Springer Handbooks. Springer, Boston, MA

[61] Agarwal U, Chandra S, Kumar V. Structural , Electronic and Elastic Properties of zincblende III-Arsenide Binary Compounds : First-Principles Study. 0(1).

[62] Pluengphon P, Bovornratanaraks T. Structural and mechanical properties of GaAs under pressure up to 200 GPa. Solid State Commun 2014;195:26–30.

[63] Güler E, Güler M. Phase transition and elasticity of gallium arsenide under pressure. Mater Res. 2014 Sep 1;17(5): 1268–72. 51. Flannery CM, Chilla E, Semenov S, Froehlich HJ. Elastic properties of GaAs obtained by inversion of laser-generated surface acoustic wave measurements. In: Proceedings of the IEEE Ultrasonics Symposium. IEEE; 1999. p. 501–4

[64] Boulder, Chapter 2: Semiconductor Fundamentals. August 2007

[65] Si F. Lecture 5 : Intrinsic Carrier concentration in intrinsic Si Carrier concentration in intrinsic semicon-. 2 (2):1–5.

[66] Surfaces S. GaAs Heterostructures and 2D electron gas. Screening. :8–16.

[67] Spirkoska D, Arbiol J, Gustafsson A, Conesa-Boj S, Glas F, Zardo I, et al. Structural and optical properties of high quality zinc-blende/wurtzite GaAs hetero-nanowires.

[68] Iskakova K, Akhmaltdinov R, Amanova A. About the energy levels of GaAs. J Phys Conf Ser. 2014;510(1).

[69] Kayali S. I. GaAs Material Properties. Kabita K, Maibam J, Sharma BI, Brojen Singh RK, Thapa RK. Density functional theory study on pressure induced structural transformation, elastic properties and electronic structure of gallium arsenide (GaAs) . Vol. 8, International Journal of Innovation and Applied Studies. 2014.

[70] Maurer RH, Chao K, Brent Bargeron C, Benson RC. RELIABILITY OF GALLIUM ARSENIDE DEVICES.

[71] Pandey BP. Electronic and Optical Properties of GaAs Armchair Nanoribbons: DFT Approach. Vol. 11, International Journal of Nanoelectronics and Materials. 2018.

Chapter 6

Comparative Analysis Carried Out on Modern Indentation Techniques for the Measurement of Mechanical Properties: A Review

Saquib Rouf, Sobura Altaf, Shezan Malik, Kaleem Ahmad Najar and M.A. Shah

Abstract

Nowadays many indentation techniques are being commonly employed for determining some mechanical properties (harness, elastic modulus, toughness, etc.) using simple method of measuring the indentation depth. On the basis of measurement of depth of penetration, indentation technique has be classified into major categories i.e. microindentation and nanoindentation. Nanoindentation technique uses indirect method of determining the contact area as the depth of penetration is measured in nanometers, while in conventional indentation the area in contact is measured by elementary measurement of the residual area after the indenter is removed from the specimen. Dynamic hardness is the best result of dynamic indentation which can be expressed as the ratio of energy consumed during a rapid indentation to the volume of indentation. The parameter which are taken into consideration are indentation depth, contact force, contact area, mean contact pressure.

Keywords: hardness, elastic modulus, nano/microindentation, Berkovich indenter, spherical indenter & vicker indenter

1. Introduction

We are having distinct techniques to rule out the mechanical properties of materials; one among them is nanoindentation. Nanoindentation is the most accepted method to regulate the mechanical properties like hardness, elastic modulus, toughness and stiffness of a material. Thus, nanoindentation is also noted by its different names in mechanical engineering field like; depth sensing indentation (DSI), instrumental indentation technique (IIT) and universal hardness test (UHT).

The indenter is the main component of nanoindentation testing which is pressured through the exterior of a material. With a prescribed load, the displacement of the indenter inside a material is observed. The hardness of the material is shown by the equation:

$$H = \left(\frac{P_{max}}{A_r}\right)$$

IntechOpen

Where, P_{max} is a maximum load for depth (h) and A_r is the residual area or the projection of indenter. This residual area inside a material is then measured by atomic force microscope. The residual area builds upon the kind of indenter and material of indenter. Therefore, the mechanical properties can vary as the penetrating material changes from one experiment to another.

The nanoindentation is also used to measure reduced elastic modulus and the sample modulus by using the following Equations [1];

$$\text{Reduced elastic modulus}, E_r = \frac{dP}{dh} \times 1/2 \times \frac{\sqrt{\pi}}{A_{r.}}$$

$$\text{Sample modulus}, E_s = \left(1 - v_s^2\right)\left(\frac{1}{E_r} - \frac{1 - v_i}{E_i}\right)$$

Where, E_i is the modulus of indenter and, v_s and v_i is the poisson's ratio for sample and indenter, respectively [1, 2].

1.1 Types of indenters

The various types of indenters with their specifications are mentioned in **Table 1** [3]. Generally, nanoindentation itself has a wide range of applications in physical science and concedes us to examine nanoscale surface adjustments in solid materials and specify the appearing variations in its mechanical response. From atomic structure to atomic defects, this has made a revolution in material engineering. It can be used to study discrete atomic rearrangement of specimen under loading conditions. When equipped with Raman spectroscopy and Atomic Force Microscope, it can find its place in studying lattice dislocations [4].

From literature survey, it has been found that results of nanoindentation depend upon the indenter tip, shape and its orientation [5]. AFM plays a vital role in nanoindentation process. It is used to measure the residual indentation area of the specimen which is further used to calculate hardness of material. It has been clearly observed and proved that nanoindentation results depend upon the type of indenter used. So, we can say that nanoindentation results vary for different indenters. To achieve better results in indentation process, continuous nanoindentation has been introduced. This approach involves intially loading the indenter at peak load and then unloading 90% of the peak load for 50 seconds and keeping it after 90% of unloading for almost 100 seconds and lastly unloading wholly the indenter. The continuous nanoindentation is used to find stiffness in terms of indentation depth in a one shot experiment.

Indenter	Projected area	Semi angle
Sphere	$2\pi Rh_p$	Not available
Berkovich	$3h_p^2(\tan\theta)^2$	65.3°
Vickers	$4h_p^2(\tan\theta)^2$	68°
Knoop	$2h_p(\tan\theta_1)(\tan\theta_2)$	θ_1 =86.25°, θ_2 = 65°
Cube Cone	$3h_p^2(\tan\theta)^2$	35.26°
Cone	$\pi h_p^2(\tan\alpha)^2$	α (effective cone angle)

Table 1.
Indenter specifications [3].

1.1.1 Spherical indenter

Spherical indenters are used for soft materials. It has been established that even a single alteration of the indenter size or radius can give collectible observation into heterogeneous aspects of the radiation-induced-damage region. The most common spherical indenter known is diamond spherical indenter which has radius less than 1 micron. We know that indentation hardness is used as; $H = \frac{P}{A}$, where, P is the applied load and A is the area of indenter.

The indentation hardness equation can be also known as $= \frac{4P}{\pi d^2}$, where, d is the diameter of contact circle when at full load. The nanoindentation deals with the size of impression, whose area is to be calculated and is found by [1–3]:

$$A = 2\pi R_i h_c$$

where, R_i is radius of the indenter and h_c is the depth of contact also called contact depth.

The Brinell hardness number for spherical indenters is calculated as [6]:

$$BHN = \frac{2P}{\pi D\left(D - \sqrt{D^2 - d^2}\right)}$$

Where, D is the diameter of indenter. The Brinell hardness number depends on the real curved surface response.

Researcher [7] has adopted the use of spherical indenter to figure out the structural characteristics of SS400, SM490 and SM520 using finite element analysis to create a shift of algorithm. The specifications taken under consideration are yield strength (σ_y), the hardening exponent (n), and the ratio (α). The α here is characterized as the ratio between the strain at beginning point of strain hardening and the yield strain. The spherical indenter with the value of 1141GPa as elastic modulus and that of poisson's ratio as 0.07, the radius of the indenter is put up to R = 5 μm and the maximum indentation depth is kept 0.3R, the results derived from the indentation process on the above mentioned specimens are very much reliable with the proposed model of reverse algorithm and can provide the best parameters of structural steel. Thus, the study of the radius of the indenter is an important factor for supposition of disorder nucleation shear stress implied from spherical indentation retortion.

Dynamic indentation of elastic plastic solid has been also reported here [8]. Dynamic hardness is the best result of dynamic indentation which can be expressed as the ratio of energy consumed during a rapid indentation to the volume of indentation. There is a restraination to either the displacement or the force of the indenter by the indentation equipment. Indentation depth, contact area, contact force, and mean contact pressure were the parameters that are taken into account and the finite element technique has been operated to determine the contact issues. The author observed that parameters like contact force, contact area and mean contact pressure are directly dependent on indentation velocity irrespective of material's elastic plastic property. The dynamic indentation of a material is linked to a dimensionless parameter which can be attained as the ratio between the kinetic energy density shifted towards the indented material to the sufficient net energy of the material and the initial yield. It has been seen that this dimensionless parameter is associated with the plastic deformation of a solid which implies larger the dimensionless parameter larger will be the plastic deformation of solid. Down below **Figure 1** distinguishes the types of indenters [9] as: (a) Spherical (b) Conical (c) Square- pyramidal (vicker) & (d) Berkovich, respectively.

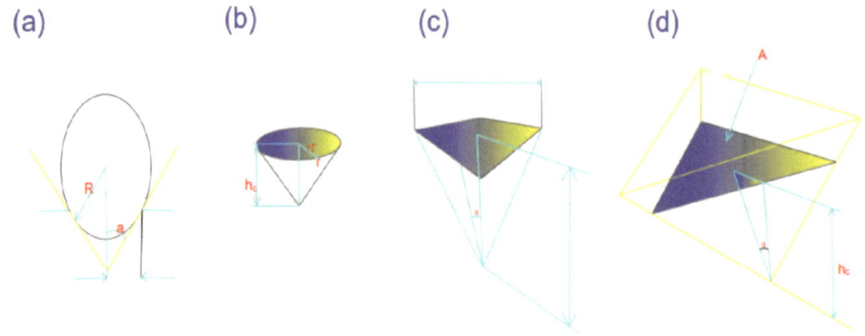

Figure 1.
Types of indenters [9]: (a) spherical (b) conical (c) square- pyramidal (vicker) & (d) Berkovich.

1.2 Berkovich indenter

The Berkovich indenter tip has a geometry of a three-sided pyramid that can be rested to a point, hence upholding a self-similar geometry to very small extent. This geometry is generally much approved than the Vickers indenter tip which has a geometry of four-sided pyramid. It is most often preferred for certaining the mechanical parameters of materials. Berkovich indenter with geometry of three sided pyramid contains a face angle of 65.3° between the sides. The indenter tip is blunted and has been constructed with the hard materials example diamond. It is more precise than vicker indenter due to its sharp point. The Berkovich indenter tip is of optimal use for most testing funtions. It is not easily impaired and can be promptly built. It brings about plasticity at slight loads too generating a relevant share of hardness. The Berkovich indenter tip is feasible as a traceable standard.

On the specimen, the projected area of Berkovich indenter is given as [3]; $A_p = 3\sqrt{3}h_c^2(\tan\theta)^2$, where θ is the face angle. For $\theta = 65.3°$, $A_P = 24.5h_c^2$.

Further, the Meyer hardness is given by [10]:

$$H = \frac{P}{A_P} = \frac{P}{24.5h_c^2}$$

From the above equation we can conclude that Meyer hardness for Berkovich indenter is the function of contact depth. Researchers [11] have reported the nanoindentation of zirconia yttria and alumina zirconi-yttria with the help of the Berkovich diamond indenters having a tip radius of 20 to 25 nm. The main function of this indentation practise is to resolve the mechanical properties like modulus of elasticity (E) and fracture toughness of the above mentioned biomedical ceramics. The values of mechanical properties are calculated at different loads like 1.5 mN, 2 mN and 5 mN. The frequency for indentation is 75 Hz with poisson's coefficient $v = 0.25$. AFM is engaged in the analysis of the continuing samples of nanoindentation. The equation given below (Sneddon equation) is employed in the calculation of the modulus of elasticity [12]:

$$S = 2\beta\sqrt{\left(\frac{A}{\pi}E_r\right)}$$

Where, A is the area of contact which in return is a function of indentation depth (h) also called as depth of penetration. E_r is reduced modulus of elasticity and β which is a constant based upon the geometry of indenter and for Berkovich indenter value of β is 1.034. The modulus of elasticity is calculated by the following Equation [13]:

$$E = \frac{1 - V^2}{\frac{1}{E_r} - \frac{1 - V_i^2}{E_i}}$$

Where, E_i, E and V_i, V are constants of modulus of elasticity and poisson's ratio of diamond indenter and specimen respectively. Further fracture toughness of the material is given as [14]:

$$K = k\left(\frac{E}{H}\right)^{\frac{1}{2}}\left(\frac{P}{C^{3/2}}\right)$$

Where, H is the hardness of material, P is the load practiced, E is the modulus of elasticity, k is empirical constant depending upon the geometry of indenter and c is the crack length. The results obtained are mentioned in **Table 2**.

Generally, Chech et al. [15] reported the bluntness of Berkovich indenter in their research. The materials that were employed were those whose young's modulus were already known; like fused silica and BK7 glass. The indentation results were compared with the results from AFM. It was found that the pointless radius of a spherical cap representing the indenter bluntness depends upon the technique accepted.

1.3 Vicker indenter

The Vickers indenter resembles to pyramid of square shape with faces and edges at angles of 68° and 148° respectively. The most common vicker indenter is vicker diamond indenter. The vicker hardness is given as [16]:

$$VH = 1.8544\frac{P}{d^2}$$

Where, d is the distance from one corner to its opposite corner of the projection left on the specimen. The Meyer hardness of vicker indenter is given by [17]:

$$H = \frac{2P}{d^2}$$

Also, the projected area of indenter, $A_p = 4h_c^2(\tan\theta)^2$, where θ is the face angle and is equal to 68°, h_c is the contact depth [18].

Material	Composition	Modulus of Elasticity (GPa)	Nano-hardness (GPa)	Stiffness (N/m)	K_{ic} (Mpa)
ATZ	20wt%Al_2O_3 + 80wt%TZ-3Y	355 ± 7	21 ± 1.2	$157{,}471 \pm 12$	4.2 ± 0.1
ZTA	80wt%Al_2O_3 + 20wt%TZ-3Y	360 ± 6	35 ± 1	$161{,}190 \pm 12$	3.50 ± 0.2
3Y-TZP	97%molZrO_2 + 3%molY_2O_3	354 ± 7	25 ± 0.8	$83{,}886 \pm 9$	5.1 ± 0.2
8Y-CSZ	92%molZrO_2 + 8%molY_2O_3	385 ± 2	31.3 ± 0.2	$153{,}423 \pm 12$	3.77 ± 0.02

Table 2.
Mechanical properties of various materials [11].

The vicker indenter is utilized in determination of the fracture toughness of brittle materials example glass. Fracture toughness (*KC*) is explained as resistance offered by a material to abrupt generation of cracks [19]. However, Anstis et al. [20] proposed an equation to measure the fracture toughness. This equation was a result of experiments on 16 materials. The equation for fracture toughness is given below:

$$K_{ic} = \chi\sqrt{\frac{E}{VH}}\frac{P}{C^{\frac{3}{2}}}$$

The above equation has an uncertainty of 25%, in this equation, *VH* is the vicker hardness, *E* is the modulus of elasticity in Mpa and χ is the dimensional less constant based upon the frame of indenter and geometry of crack generated. Usually χ varies from 0.016 to 0.004 for vicker indenter.

Herval et al. [21] reported the comparison of fracture toughness from vicker indenter using Anstis equation for these materials; crown glass (BK7), heavy flint glass (SF17), zerodur glass and hydroxyapatite ceramic. The experimentation has been performed on Zwick Roell with load varying from 2 to 100 N with dwell time at maximum load of 10 seconds. The Cracks are examined using optical microscope. Thus, the results obtained are summarized in the **Table 3**. The authors conclude by verifying that fracture toughness remains constant for a specific material under varying loads. This means that polishing does not generate residual stress on sample.

1.3.1 Fracture toughness of Y-TZP dental ceramic

Fracture toughness is the basic criteria for studying the potential of bio ceramics. The most common ceramic used is zirconia. It finds its application in bone and dental implants. Y-TZP dental ceramic contains 2–3 mol% of yttrium oxide which produces of 6MPam1/2 [22]. Donaka [23] reported the indentation of Y-TZP dental ceramic by vicker indenter. The samples were cut into 10 × 10 × 2 mm plates. Following four set of loads were applied; 24.03 N (VH3), 49.03 N (VH5), 196.13 N (VH20) and 294.20 N (VH30). Each load has been applied 30 times. The fracture toughness has been directly determined by crack length. **Table 4** can sum up the above experiment.

The data above shows that with the increase in applied load, the hardness of Y-TZP increases. This reaction is recognized as normal indentation size effect [24]. Fracture toughness depends upon the type of crack formed. So, for the process of attaining the fracture toughness of a material, the fundamental step is to notice the type of crack formed. Usually palmqvist and median cracks are formed due to vicker indentation. Palmqvist cracks can be found in tough materials at high loads and in brittle materials at low loads [25]. Depending upon the type of crack developed various methods have been used to calculate fracture toughness. Some of them are mentioned in the **Table 5**.

The vicker indentation on Y-TZP Ceramic results in crack propagation which is further used to calculate fracture toughness. Crack propagation mainly depends

Parameter	Crown glass	Heavy flint glass	Zerodur glass	Hydroxyapatite ceramic
K_{ic} (MPa)	0.5±0.40	0.47 ± 0.03	0.93 ± 0.1	1.20 ± 0.03
VH (GPa)	6.3 ± 0.3	4.4 ± 0.3	6.7 ± 0.6	4.4 ± 0.6

Table 3.
Comparison of fracture toughness by vicker indenter.

Load	Vicker hardness (average)
VH3	1379
VH5	1344
VH20	1345
VH30	1337

Table 4.
Fracture toughness of Y-TZP with varying loads.

Author	Crack type	Equation
Casellas [26]	Palmqvist	$0.024\frac{F}{C^{3/2}}\left(\frac{F}{VH}\right)^{1/2}$
Palmqvist [27]	Palmqvist	$0.0028VH^{1/2}.\left(\frac{F}{VH}\right)^{1/2}$
Shetty et al. [28]	Palmqvist	$0.0319\frac{F}{al^{1/2}}$
Niihara et al. [29]	Palmqvist	$0.0089\left(\frac{E}{VH}\right)^{2/5}\frac{F}{al^{1/2}}$
Anstis [30]	Median	$0.016\frac{F}{C^{3/2}}\left(\frac{E}{VH}\right)^{1/2}$
Evans and Charles [30]	Median	$0.0752\frac{F}{C^{3/2}}$
Tanaka [30]	Median	$0.0725\frac{F}{C^{3/2}}$
Niihara, Morena and Hasselman [29]	Median	$0.0309\left(\frac{E}{VH}\right)^{2/5}\frac{F}{C^{3/2}}$
Lankford [28]	Anykind	$0.0782(VHa^{1/2})\left(\frac{E}{VH}\right)^{2/5}\left(\frac{C}{a}\right)^{1.56}$

Table 5.
Various methods to calculate fracture toughness depending upon the crack formed [26–30].

upon the indentation load applied and the type of material. For E = 210 GPa, the fracture toughness of palmqvist crack varies from 4.96 to 7.73 MPam1/2 and 3.96 to 6.72 MPam1/2 for median crack profile. The lankford model gives the highest of values 7.73 MPam1/2 for 29.42 N and Anstis give lowest of 4.49MPam1/2 for 294.20 [28–30].

1.4 Nanoindentation for nuclear materials

The nanoindentation technique for irradiated materials started back in 1986 [31]. From last few decades, nanoindentation has been acknowledged as well founded means to explore the bounded mechanical properties at small scale. Therefore, the idea of the sequence of ion irradiation and nanoindentation has been greatly advanced in later years to examine the mechanical role of nuclearly arranged materials with irradiation developed.

Various studies [32–34] have been performed on the different combination of materials with the focus on effect of crystal structure and irradiation temperature on hardness of material. For ion irradiation there is a plan of action that needs to be followed to observe the irradiation damage prompt by energetic neutrons. It is very difficult to evaluate the mechanical properties of ion irradiated materials. The most important factors are limited indentation depths and inhomogenous distribution of irradiation induced defects [35]. Additionally, upon ion-irradiation the metal surface is modified by a thin radiation-damaged layer which causes a change in its mechanical response as compared to the bulk of the sample. In order to successfully study the effects of radiation damage on the indentation behavior, we need to first decouple the effects of orientation from the effects of the increased defect density caused by irradiation. It was observed that there exists a strong orientation effect of radiation which induces mechanical changes at the grain scale. Obviously, surface energies for the orientation is quite different. So there can be significant differences in the damage experienced by these grains [36, 37].

The impact of high energy particles on the metalic materials for energy systems are widely studied as the mechanical properties of these materials are highly effected [38, 39]. This study is very important to develop a smart and effective nuclear energy production system. Mechanical characterization methods like uniaxial compression and tension remain invalid [40], as high spatial resolution is required [41–43]. Since nanoindentation gives rich statistical data [44, 45], hence it is convenient for mechanical characterization of ion irradiated materials. Nanoindentation for irradiation material can be broadly classified into two categories; suface nanoindendation and cross-sectional nanoindentation [36, 46].

For irradiated materials, spherical indentation is widely preferred over berkovich indentation as it can analyze the irradiation effect not only on materials yielding but also the elasto-plastic transition and strain hardening behavior, by converting the force-depth (F h) relationships into the indentation stress–strain (ISS) relationships [47, 48]. High temperature nanoindentation can be used to determine the thermo-mechanical behavior of irradiated materials at the actual working temperature [49].

1.5 Nanoindentation for polymeric materials

From past one decade, the use of polymeric composite matrix has been widely used due to their high energy absorption, light weight and good adhesion [50]. A large number of researchers [50–53] have reported the nanoindentation on polymeric surfaces and it has been observed that assessment of elastic modulus has remained a challenge. For nanoindentation of soft polymeric materials, the materials complaince can create difficulties in tracing the initial contact [54]. Researcher [55] has shown the main problem of nanoindentation with polymeric materials, the elastic moduli obtained by nanoindentation does not coincide with the moduli obatined by conventional micro tension and compression. There has been an increase in the moduli obtained by the nanoindentation of polymeric materials. Studies performed on polystyrene and polycarbonate show 64% and 70% shift in moduli, respectively [52]. Moreover, for PMMA [56], 67% shift is seen and 20% for poly bemzocyclobutene [53]. The elastic moduli for nanoindentation have been plotted against the micro moduli of the selected set of polymers, shown in **Figure 2(a)**.

The other approaches like dynamic modulus approach are applied. This method is often termed as continuous stiffness measurement. The two important parameters in this are storage moduli and loss moduli [57]. The parameters can be also used with dynamic mechanical analysis (DMA). The storage moduli and loss moduli were obtained by both dynamic indentation and DMA for thermoplastic materials [58]. The results of the two methods were compared with each other as shown in **Figure 2(b)**. The consistent trend of greater dynamic indentation moduli than DMA moduli was observed for various materials. By this study, it can be observed that dynamic indentation moduli's can be subjected to a percent of error while characterizing the polymeric materials [59].

1.6 Nanoindentation in biological materials

The mechanical properties of biological materials are the early stage of development. Among all the techniques for mechanical characterization of biological

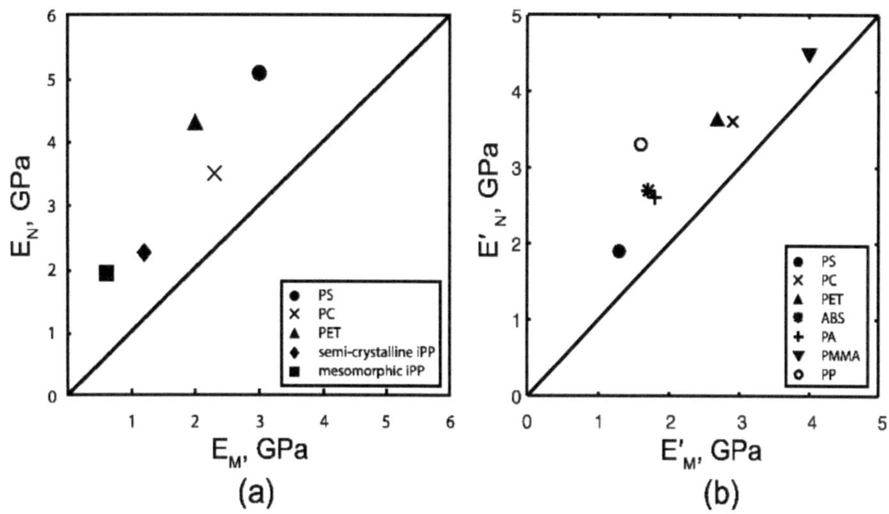

Figure 2.
(a) Surface nanoindentation and (b) cross-sectional nanoindentation [46].

materials, nanoindentation is most powerful tool due to its wide nanolevel force resolution [60] and mainly because of its surface and interfacial properties which play significant role not only in biology but also in synthesis of biomaterials. Therefore, the study of adoption of nanoindentation for biological materials and systems is of great interest.

New studies have been performed were it has been described how the diseases respond towards the mechanical characteristics up to the molecular level [61–66]. The method has also been used to carry out the mechanical analysis of fossils [67]. Indentation finds its application in describing the behavior of human skin which can be used to solve the aging problem of skin [68, 69]. Application of nanoindentation technique to biomaterials research is also expanded up to human enamel and eye tissues. In the coming years, the importance of nanoindentation testing in biology and biomaterials research is likely to show a rapid increase. Nanoindentation testing is implicated on both soft and hard tissues and also on biomaterials especially those with hierarchical microstructures. The nanoindentation methods are also applied to highlight the study of skeletal and dental tissues [70]. There also exist nanomechanics for soft, living tissues and polymeric biomaterials. Nanoindentation is now also being applied to the problem with studies examining connective tissues and polymers using both static and dynamic methods. The nanoindentation investigations on natural biomaterials have contributed significantly to biomimetics and the development of new composite materials. Continuing advancements in nanoindentation analysis will increase the method's utility in the characterization of biomaterials [71].

The measurement protocol for nanoindentation of biological samples (sometimes referred as bioindentation) takes into account the irregularity of the surface of the samples by incorporating an automatic surface detection procedure in the measurement matrix. The use of large spherical indenters facilitates contact detection on extremely soft samples (hydrogels [72], cartilage, scaffolds) by providing larger contact stiffness and averages surface and structural inhomogeneity. The penetrations seen in bioindentation are usually in the range of ten to several hundred micrometers, thus testing a large volume of tissue rather than single cells [73]. The applications of the bioindenter are very wide. Many human tissues are subject to mechanical loading and their mechanical characterization can provide valuable information for disease evolutions, treatments and also developments of artificial replacements (implants, scaffolds). Bioindentation can be used in the diagnostic of disease (liver functions, arteries) and for fundamental research on treatment of these diseases [74].

2. Conclusion

Nanoindentation is a dynamic perceptible method for attaining mechanical properties from very limited content. In delicately regulated tests in which the acceptance of the elastic contact analysis are met, accuracy of a few percent is smoothly obtainable for indentations as micro as 10 nm. Specialists must be constantly aware of the holdings of variations from these suppositions on nanoindentation results. Exact evaluation for load, displacement and machine concurrence are requirement, as is an effective rational sketch of the shape of the tip, and a configuration devised to reduce the consequences of thermal drift and plasticity.

If there arise a need to measure a surface layer or other small volume, then evading the effects of surface conditions, or nearby free surfaces or interfaces, a practical indentation size range can be found analytically. Size effects, pile-up and anisotropy can advance precise deviations that must be alleged.

Nanoindentation is an area of powerful research and development and data analysis practices are frequently being revised. Future developments of nanoindentation are foreseen in the area of periodic incorporation of computer simulations, (FEM and others) and quantitative imaging technique to boost in analysis of depth-sensing indentation data, and much development in dynamic contact analysis system in addition to incorporation of acoustic practices.

Author details

Saquib Rouf[1*], Sobura Altaf[1], Shezan Malik[1], Kaleem Ahmad Najar[2] and M.A. Shah[3]

1 SSM College of Engineering and Technology, Baramulla, India

2 Islamic University of Science and Technology, Awantipora, India

3 Department of Physics, National Institute of Technology-Srinagar, J&K, India

*Address all correspondence to: khansaquib626@gmail.com

References

[1] W.C Oliver, G.M Pharr (1992) An improved technique for determining hardness and elastic modulus using load and displacement sensing indentation experiments. Materials Research Innovation 7:1564.

[2] A. Tiwari, S. Natrajan (2017) Applied Nanoindentation in Advanced Materials. John Willey. ISBN: 978111908450.

[3] Anthony C, Fischer – Cripps (2011) Nanoindentation, Third Edition. Springer. ISBN: 978–1–4419-9871–2.

[4] Christopher A. Schuh (2006) Nanoindentation Studies of Materials. Materials Today 9(5): 32–40.

[5] Sagadevan S, Murugasen P (2014) Novel Analysis on the Influence of Tip Radius and Shape of the Nanoindenter on the Hardness of Materials. Procedia Materials Science (6): 1871–1878.

[6] G. Sundaranjan,M. Roy (2001) Hardness Testing. Encyclopedia of Materials: Science and Technology (Second Edition. PP: 3728–3736.

[7] T.H Pham, Q.M Phan and S.E Kim (2018) Identification of The Plastic Properties of Structural Steel Using Spherical Indentation. Materials Science and Engineering 711: 44–61.

[8] A. Lee, K. Komvopoulos (2018) Dynamic Spherical Indentation of Elastic- Plastic Solids. International Journal of Solids and Structures 146: 180–191.

[9] A.N. Masir, A. Darvizeh, A. zajkani (2019) Nanoindentation on Bio Inspired High Performance Nature Composite by Molecular Dynamic Method. Advanced Composites Letters 28:1–13.

[10] E. Broitman (2017) Indentation Hardness Measurement at Macro-, Micro-, and Nanoscale: A Critical Overview. Tribology Letters 65:23

[11] M.C. Aragon-Durate, A. Nevarez-Rascon, H.E. Esparza-Ponce, M. M. Nevarcz Rascon, R. P. Talamantes, C. Ornelas, J. Mendez-Nonell, J. Gonzalez-Hermandez, M. J. Yacaman, A. Hurtado-Macias (2017) Nanomechanical Properties of Zirconia-Yttria and Alumina Zirconia- Yttria Biomedical Ceramics, Subjected to Low Temperature Aging. Ceramics International 43(5):3931–3939

[12] I.N.Sneddon (1965) The Relation Between Load and Penetration in The Axisymmetric Boussineq Problem for a Punch of Arbitrary Profile. International Journal of Engineering Science 3(1):47–58.

[13] F. Doerner, D. S. Gardner, W. D. Nix (1986) Plastic Properties of Thin Films on Substrates as Measured by Submicron Indentation Hardness and Substrate Curvature Techniques. Journal of Materials Research 1(6): 845–851.

[14] H. Hertz (1881) On The Contact of Rigid Elastic Solids. J. Reine A19ngew. Math. 92, 1881, pp. 156–171.

[15] J. Cech, P. Hausild, O. Koravik, A. Materna (2016) Examination of Berkovich Indenter Tip Bluntness. Materials & Design 109: 347–357

[16] G.B. Ghorbal, A. Tricoteaux, A. Thuault, G. Louis, D. Chicot (2017) Comparison of Conventional Knoop and Vickers Hardness of Ceramic Materials. Journal of the European Ceramic Society37(6):2531–2535

[17] J. Gubicza, A. Juhasz, J. Lendvai (1996) A New Method for Hardness Determination from Depth Sensing Indentation Tests. Journal of Materials Research, 11(12):2964–2967.

[18] Y. Lu, Y. Su, W. Ge, T. Yang, Z. Yan, Y. Wang, S. Xia (2020) Conversion Between Vickers Hardness and Nanohardness by Correcting Projected Area with Sink-in and Pile-up Effects. Plasma Science and Technology 22 (06) 5602.

[19] M. Buijs, K. Korpel-Van Houten (1993) A Model for Lapping of Glass. Journal of Materials Science (28):3014–3020.

[20] G.R. Anstis, P. Chantikul, B. R lawn, D. B Marshall (1981) A Critical Evaluation of Indentation Techniques for Measuring Fracture Toughness: I, Direct Crack Measurements. Journal of the American Ceramic Society 64(9): 533–538.

[21] I. Hervas, A. Montagne, A. Van Gorp, M. Bentoumi, A. Thuault, A. Iost (2016) Fracture Toughness of Glasses and Hydroxyapatite: A Comparative Study of 7 Methods by Using Vickers Indenter. Ceramic International 42(11): 12740–12750

[22] J. Chevalier, L. Gremillard, S. Deville (2007) Low-Temperature Degradation of Zirconia and Implications for Biomedical Implants. Annual Review of Materials Research 37: 1–32.

[23] D. Coric, M. Majc Renjo, L. Curkovic (2017) Vicker Indentation Fracture Toughness of Y-TZP Dental Ceramics. International Journal of Refractory Metals and Hard Materials 64:14–19.

[24] M. Majic Renjo, L. Curkovic, S. Stefancic, D. Coric (2014) Indentation Size Effect of Y-TZP Dental Ceramics. Dental Materials 30(12):371–376.

[25] K. Niihara, R. Morena, D. P. H. Hasselman (1982) Evaluation of K Ic of Brittle Solids by the Indentation Method with Low Crack-to-Indent Ratios. Journal of Materials Science Letters 1(1): 13–16.

[26] D. Cesellas, M.M. Nagl, L. Llanes, M. Anglada (1996) Growth of Small Surface Indentation Cracks in Alumina and Zirconia Toughened Alumina. Key Engineering Materials 127–131:895–902.

[27] S. Palmqvist (1957) A Method to Determine The Fracture Toughness Brittle Materials, Especially Hard Metals. Jenkontorets Ann 141:303–307.

[28] S. Kaur, R.A. Cutler, D.K. Shetty (2009) Short-Crack Fracture Toughness of Silicon Carbide. Journal of the American Ceramic Society 92(1): 179–185.

[29] K. Niihara, R. Morena, D. P. H. Hasselman (1982) Evaluation of K Ic of Brittle Solids by the Indentation Method with Low Crack-to-Indent Ratios. Journal of Materials Science Letters 1(1): 13–16.

[30] A. Nastic, A. Merati, M. Bielawski, M. Bolduc, O. Fakolujo, M. Nganbe (2015) Instrumented and Vickers Indentation for the Characterization of Stiffness, Hardness and Toughness of Zirconia Toughened Al2O3 and SiC Armor. Journal of Material Science and Technology 31(8):773–783.

[31] S. J. Zinkle, W. C. Oliver (1986) Mechanical Property Measurements on Ion-Irradiated Copper and Cu-Zr. Journal of Nuclear Materials 141–143 (1): 548–552.

[32] K. Sato, H. Yamashita, A. Hirosako, S. Komazaki, Q. Xu, M. Onoue, R. Kasada, K. Yabuuchi, A. Kimura (2018) Investigation of Mechanical Properties of Stress- Relieved and Electron – Irradiated Tungsten after Hydrogen Charging. Nuclear Materials and Energy 17:29–33.

[33] B. Su, D.Y.W. Yu, H. Liang, G. Liu, Z. Huang, X. Liu, Z. Chen (2017) Damage Development of Sintered SiC Ceramics with the Depth Variation in Ar Ion-Irradiation at 600 °C. Journal of

European Ceramic Society 38(5): 2289–2296.

[34] Y. Chen, Y. Liu, E. G. Fu, C. Sun, K. Y. Yu, M. Song, J. Li, Y. Q. Wang, H. Wang, X. Zhang (2015) Usual Size – Dependent Mechanisms in Helium Ion – Irradiated Immiscible Coherent Cu/Co Nanolayers. Acta Materialia 84: 393–404.

[35] X. Xiao, L. Yu (2020) Nano-Indentation of Ion- Irradiated Nuclear Structural Materials: A Review. Nuclear Materials and Energy 22: 100721.

[36] Q.M. Wei, N. Li, *N. Mara*, M. Nastasi, A. Misra (2011) Suppression of Irradiation Hardening in Nanoscale V/Ag Multilayers. Acta Materialia 59(16): 6331–6340.

[37] P. Hosemann, J.G. Swadener, D. Kiener, G.S. Was, S. A Maloy, N.Li (2008) An Exploratory Study to Determine Applicability of Nano – Hardness and Micro – Compression Measurements for Yield Stress Estimation. Journal of Nuclear Materials 375(1): 135–143.

[38] I.J. Beyerlein, M.J. Demkowicz, A. Misra, B.P. Uberuaga (2015) Defect – Interface Interactions. Progress in Materials Science 74:125–210.

[39] X. Zhang, K. Hattar, Y. Chen, L. Shao, J. Li, C. Sun, K. Yu, Nan Li, M. L. Taheri, H. Wang, J. Wang, M. Nastasi (2018) Radiation Damage in Nanostructured Materials 92: 217–321.

[40] C. Shin, H. Jin, M. Kim (2009) Evaluation of the Depth-Dependent Yield Strength of a Nanoindentation Ion – Irradiated Fe-Cr Model Alloy by Using a Finite Element Modeling. Journal of Nuclear Materials 392(3): 476–481.

[41] D.E.J. Armstrong, C.D. Hardie, J.S. K.L. Gibson, A.J. Bushby, P.D. Edmondson, S.G. Roberts (2015) Small – Scale Characterisation of Irradiated Nuclear Material: Part II Nanoindentation and Micro- Cantilever Testing of Ion Irradiated Nuclear Materials 462: 374–381.

[42] J. A. Sharon, K. Hattar, B.L. Boyce, L.N. Brewer (2013) Compressive Properties of <110> Cu Micro-Pillars after High – Dose Self-Ion Irradiation. Materials Research Letters 2(2): 57–62.

[43] T. Wei, A. Xu, H. Zhu, M. Lonescu, D. Bhattacharyya (2017) In Situ Micro-Compression Testing of He^{2+} Ion Irradiated Titanium Aluminide. Nuclear Instruments and Methods in Physics Research Section B: Beam Interactions with Materials and Atoms 409: 288–292.

[44] M. Saleh, A. Xu, C. Hurt, M. Lonescu, J. Daniels, P. Munroe, L. Edwards, D. Bhattacharyya (2019) Oblique Cross - Section Nano Indentation for Determining the Hardness Change in Ion- Irradiated Steel. International Journal of Plasticity 112: 242–256.

[45] N. K. Mukhopadhyay, P. Paufler (2013) Micro- and Nanoindentation Techniques for Mechanical Characterisation of Materials. International Materials Reviews 51 (4): 2019–245.

[46] D. Kiener, A.M. Minor, O. Anderoglu, Y. Wang (2012) Application of Small – Scale Testing for Investigation of Ion – Beam -Irradiated Materials. Journal of Materials Research 27(21):2724–2736.

[47] J. Weaver, C. Sun, Y. Wang, S.R. Kalidindi, R.P. Doerner, N.A. Mara, S. Pathak (2018) Quantifying the Mechanical Effects of He, W and He+W Ion Irradiation on Tungsten with Spherical Nanoindentation. Journal of Material Science 53: 5296–5316.

[48] S. Pathak, S.R. Kalidindi, J.S. Weaver, Y. Wang, R.P. Doerner, N.A. Mara (2017) Probing Nanoscale Damage

Gradients in Ion-Irradiated Metals Using Spherical Nanoindentation 7: 11918.

[49] Q. Dong, H. Qin, Z. Yao, M.R. Daymond (2019) Irradiation Damage and Hardening in Pure Zr and Zr-Nb Alloys at 573 K from Self-Ion Irradiation. Materials & Design 161: 147–159.

[50] H. Hardiman, T.J. Vaughan, C.T. McCarthy (2017) A Review of Key Developments and Pertinent Issues in Nanoindentation Testing of Fibre Reinforced Plastic Microstructures. Composite Structures 180:782–792.

[51] D. Tranchida, S. Piccarolo, J. Loos, A. Alexeev (2006) Accurately evaluating Young's modulus of polymers through nano indentations: A phenomenological correction factor to the Oliver and Pharr procedure. Applied Physics Letters 89: 171905.

[52] D. Tranchida, S. Piccarolo, J. Loos, A. Alexeev (2007) Mechanical Characterization of Polymers on a Nanometer Scale through Nano indentation. A Study on Pile - up and Viscoelasticity. Macromolecules 40: 1259–1267.

[53] M. R. Vanlandingham, J. S. Villarrubia, W. F. Guthrie, G. F. Meyers (2001) Nanoindentation of Polymers: An Overview. Macromolecular Symposia 167(1):15–44.

[54] J. Deuschle (2008) Mechanics of soft polymer identification. Universitat Stuttgart: http://dx.doi.org/ 10.18419 / opus-907

[55] J. M. Kranenburg, C.A. Tweedie, K. J. V. Vliet, U. S. Schubert (2009) Challenges and Progress in High-Throughput Screening of Polymer Mechanical Properties by Indentation. Advanced Materials 21(35):3551–3561.

[56] H. Lu, B. Wang, J. Ma, G. Huang, H. Viswanathan (2003) Measurement of Creep Compliance of Solid Polymers by Nanoindentation. Mechanics of Time-Dependent Materials 7: 189 207.

[57] G. M. Odegard, T.S. Gates, H. M. Herring (2005) Characterization of Viscoelastic Properties of Polymeric Materials Through Nanoindentation. Experimental Mechanics 45: 130–136.

[58] P. Frontini, S. Lotfian, M. A. Monclus, J. M. Aldareguia (2015) High Temperature Nanoindentation Response of RTM6 Epoxy Resin at Different Strain Rates. Experimental Mechanics 55(5):1–12.

[59] J. Giro-Paloma, J. J. Roa, A. M. Diez-Pascual, E. Rayon, *A. Flores*, M. Martínez, J. M. Chimenos, A.I. Fernandez (2013) Depth-Sensing Indentation Applied to Polymers: A Comparison Between Standard Methods of Analysis in Relation to the Nature of the Materials. European Polymer Journal 49(12): 4047–4053.

[60] N. E. Kurland, Z. Drira, V. K. Yadavalli (2012) Measurement of Nanomechanical Properties of Biomolecules Using Atomic Force Microscopy. Micron 43 (2–3): 116–128.

[61] S. Gracia-Manyes, O. Domenech, F. Sanz, M. T. Montero, J. Harnandez-Borrell (2007) Atomic Force Microscopy and Force Spectroscopy Study of Langmuir–Blodgett Films Formed by Heteroacid Phospholipids of Biological Interest. Biochimica et Biophysica Acta (BBA) – Biomembranes 1768(5):1190–1198.

[62] S. Huang, D. E. Ingber (2005) Cell Tension, Matrix Mechanics, and Cancer Development. Cell Cancer 8(3):175–176.

[63] D. E. Discher, P. Janmey, Y. Wamg (2005) Tissue Cells Feel and Respond to the Stiffness of Their Substrate. Science 310(5751): 1139–1143.

[64] A. J. Engler, S. Sen, H. L. Sweeney, D. E. Discher (2006) Matrix Elasticity

Directs Stem Cell Lineage Specification. Cell 126 (4): 677–689.

[65] M. Stolz, R. Gottardi, R. Raiteri, S. Miot, I. Martin, R. Imer, U. Staufer, A. Randucanu, M. Duggelin, W. Baschong, A.U. Daniels, N. F. Friederich, A. Aszodi, U. Aebi (2009) Early Detection of Aging Cartilage and Osteoarthritis in Mice and Patient Samples Using Atomic Force Microscopy. Nature Nanotechnology 4:186–192.

[66] R. J. H. Cloots, J. A. W. van Dommelen, S. Kleiven, M. G. D. Geers (2012) Multi-scale Mechanics of Traumatic Brain Injury: Predicting Axonal Strains from Head Loads. Biomechanics and Modeling in Mechanobiology (12): 137–150.

[67] S. E. Olesiak, M. Sponheimer, J. J. Eberle, M. L. Oyen, V. L. Ferguson (2010) Nanomechanical Properties of Modern and Fossil Bone. Palaeogeography, Palaeoclimatology, Palaeoecology 289 (1–4): 25–32.

[68] H. Oxlund, J. Manschot, A. Viidik (1988) The Role of Elastin in the Mechanical Properties of Skin. Journal of Biomechanics 21(3): 213–218.

[69] A. Agache, P. Humbert (2004) Measuring the Skin. Springer-Verlag Berlin Heidelberg. ISBN: 978–3–642-05691-8.

[70] K. Subramani, W. Ahmed (2017) Emerging Nanotechnologies in Dentistry-2nd Edition. Micro & Nanotechnologies Series Elsevier, ISBN: 978–0–12-812291-4.

[71] A.B. Mann (2005) Nanoindentation - Surfaces and Interfaces for Biomaterials, Editor(s): P. Vadgama. Woodhead Publishing Series in Biomaterials (225–247). ISBN: 978–1–855-73930-7.

[72] Y. Hu, J. You, D. Auguste, Z. Suo, J. Vlassak (2012) Indentation: A simple, Non-destructive Method for Characterizing the Mechanical and Transport Properties of pH-Sensitive Hydrogels. Journal of Materials Research 27(1): 152–160.

[73] C. Cai, Q. Yu, W. Li, J. Zheng, Z. Zhou (2017) Experimental Creep Behavior of Porcine Liver under Indentation with Laparoscopic Grasper for MIS Applications. Biosurface and Biotribology 3(2): 56–65.

[74] J. Nohava, M. Swain, P. Eberwein (2014) Micromechanical Properties of Polyacrylamide Hydrogels Measured by Spherical Nanoindentation. Key Engineering Materials 606: 121–124.

Chapter 7

High-Efficiency GaAs-Based Solar Cells

Masafumi Yamaguchi

Abstract

The III-V compound solar cells represented by GaAs solar cells have contributed as space and concentrator solar cells and are important as sub-cells for multi-junction solar cells. This chapter reviews progress in III-V compound single-junction solar cells such as GaAs, InP, AlGaAs and InGaP cells. Especially, GaAs solar cells have shown 29.1% under 1-sun, highest ever reported for single-junction solar cells. In addition, analytical results for non-radiative recombination and resistance losses in III-V compound solar cells are shown by considering fundamentals for major losses in III-V compound materials and solar cells. Because the limiting efficiency of single-junction solar cells is 30-32%, multi-junction junction solar cells have been developed and InGaP/GaAs based 3-junction solar cells are widely used in space. Recently, highest efficiencies of 39.1% under 1-sun and 47.2% under concentration have been demonstrated with 6-junction solar cells. This chapter also reviews progress in III-V compound multi-junction solar cells and key issues for realizing high-efficiency multi-junction cells.

Keywords: solar cells, GaAs, InP, InGaP, III-V compounds, multi-junction, tandem, high efficiency, radiation-resistance

1. Introduction

The III-V compound solar cells represented by GaAs solar cells have advantages such as high-efficiency potential, possibility of thin-films, good temperature coefficient, radiation-resistance and potential of multi-junction application compared crystalline Si solar cells. The III-V compound solar cells have contributed as space and concentrator solar cells and are important as sub-cells for multi-junction solar cells. As a result of research and development, high-efficiencies [1, 2] have been demonstrated with III-V compound single-junction solar cells: 29.1% for GaAs, 24.2% for InP, 16.6% for AlGaAs, and 22% for InGaP solar cells. **Figure 1** shows historical record-efficiency of GaAs, InP, AlGaAs and InGaP single-junction solar cells along with their extrapolations [3].

The data can be fitted with the Goetzberger function [4]:

$$\eta(t) = \eta_{limit}[1 - \exp.(t_0 - t)/c], \quad (1)$$

where $\eta(t)$ is the time-dependent efficiency, η_{limit} is the practical limiting efficiency, t_0 is the year for which $\eta(t)$ is zero, t is the calendar year, and c is a characteristic development time. Fitting of the curve was done with three parameters which are given in **Table 1**. The extrapolations show that the progress of

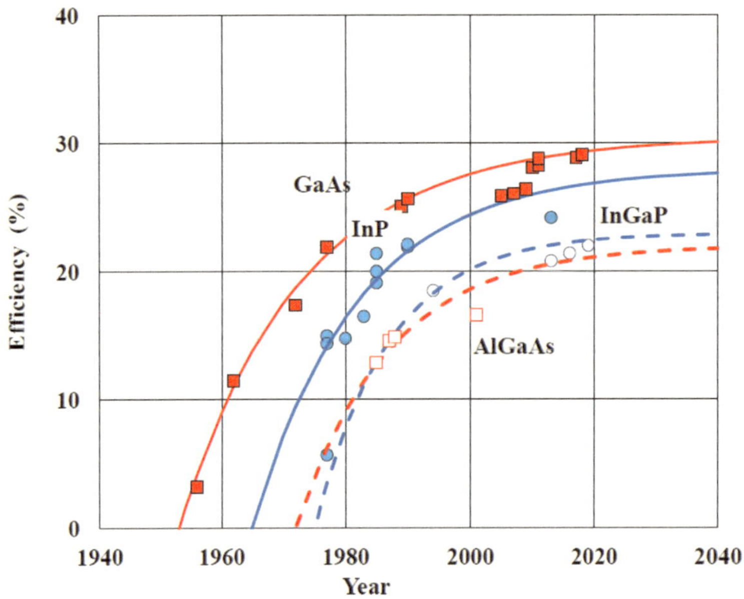

Figure 1.
World record efficiencies of GaAs, InP, AlGaAs and InGaP single-junction solar cells over years. Solid and dashed lines are the fitted trajectories using Eq. (1).

Solar cells	η_{limit}	c	t_0
GaAs	30.5	20	1953
InP	28	17	1965
AlGaAs	22	15	1972
InGaP	23	12	1975

Table 1.
Fitting parameters for various solar cells.

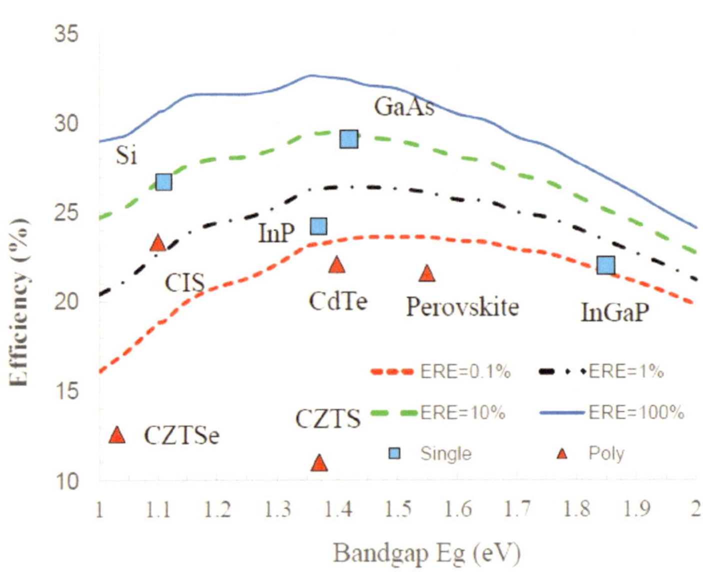

Figure 2.
Calculated and obtained efficiencies of single-junction single-crystalline and polycrystalline solar cells.

efficiencies is converging or will converge soon, which is mainly bounded by the Shockley-Queisser limit [5].

Figure 2 shows calculated and obtained efficiencies of single-junction single-crystalline and polycrystalline solar cells [6]. Because the limiting efficiency of single-junction solar cells is 30-32% as shown in **Figure 2**, multi-junction solar cells have been developed and InGaP/GaAs based 3-junction solar cells are widely used in space. Recently, highest efficiencies of 39.2% under 1-sun and 47.1% under concentration have been demonstrated with 6-junction solar cells [7].

This Chapter reviews progress in III-V compound single-junction solar cells such as GaAs, InP, AlGaAs and InGaP cells. In addition, analytical results for non-radiative recombination and resistance losses in III-V compound solar cells by considering fundamentals for major losses in III-V compound materials and solar cells. This chapter also reviews progress in III-V compound multi-junction solar cells and key issues for realizing high-efficiency multi-junction cells.

2. Analysis of non-radiative recombination and resistance losses of single-junction solar cells

By using our analytical model [8, 9], potential efficiencies of various solar cells are discussed. This model considers the efficiency loss such as non-radiative recombination and resistance losses, which are reasonable assumption because conventional solar cells often have a minimal optical loss. The non-radiative recombination loss is characterized by external radiative efficiency (ERE), which is the ratio of radiatively recombined carriers against all recombined carriers. In other words, we have ERE = 1 at Shockley-Queisser limit [5]. EREs of state-of-the-art solar cells can be found in some publications such as references [2, 10–13]. In this chapter, the EREs of various solar cells are estimated by the following relation [14]:

$$V_{oc} = V_{oc:rad} + (kT/q)\ln(ERE), \qquad (2)$$

where V_{oc} the measured open-circuit voltage, k the Boltzmann constant, T the temperature, and q the elementary charge. $V_{oc:rad}$ the radiative open-circuit voltage and is expressed by the following Eq. [15]

$$V_{oc,rad} = (kT/q)\ln\left[J_{ph}\right]V_{oc,rad})/J_{o,rad} + 1], \qquad (3)$$

where $[J_{ph}]V_{oc,rad}$ is the photocurrent at open-circuit in the case when there is only radiative recombination and $J_{o,rad}$ the saturation current density in the case of radiative recombination.

0.28 V for E_g/q - $V_{oc;rad}$ value reported in [15–17] were used in our analysis. Where E_g is the bandgap energy. The second term on the right-hand side of Eq. (2) is denoted as $V_{oc;nrad}$, the voltage-loss due to non-radiative recombination and is expressed by the following Eq. [15].

$$\Delta V_{oc;\,nrad} = V_{oc:rad} - V_{oc} = (kT/q)\ln\,[J_{rad}(V_{0c})/J_{rec}(V_{oc})] = -(kT/q)\ln(ERE), \qquad (4)$$

where $J_{rad}(V_{0c})$ is the radiative recombination current density and $J_{rec}(V_{oc})$ is the non-radiative recombination current density.

Figure 3 shows open-circuit voltage drop compared to band gap energy ($E_g/q - V_{oc}$) and non-radiative voltage loss ($V_{oc,nrad}$) in GaAs, InP, AlGaAs and

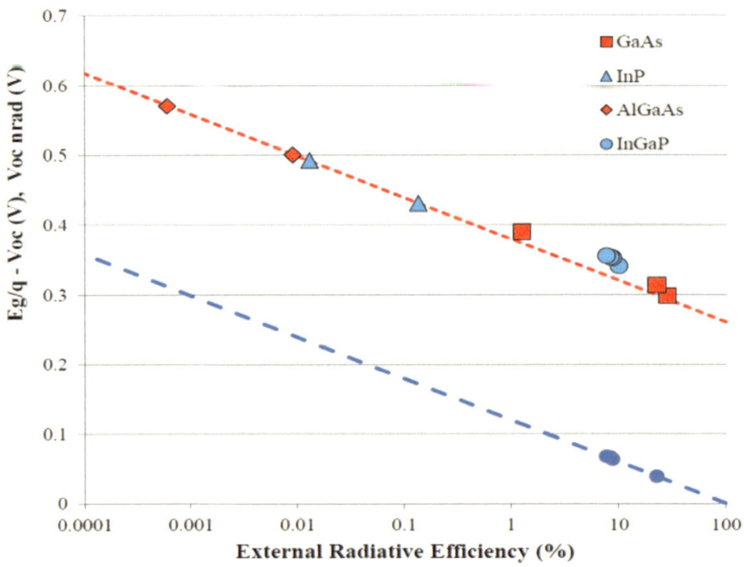

Figure 3.
Open-circuit voltage drop compared to band gap energy ($E_g/q - V_{oc}$) and non-radiative voltage loss ($V_{oc,nrad}$) in GaAs, InP, AlGaAs and InGaP solar cells as a function of ERE.

InGaP solar cells [2, 8–13, 17] as a function of ERE. High ERE values of 22.5% and 8.7% have been observed for GaAs and InGaP, respectively compared to InP (0.1%) and AlGaAs (0.01%).

The resistance loss of a solar cell is estimated solely from the measured fill factor. The ideal fill factor FF_0, defined as the fill factor without any resistance loss, is estimated by [18].

$$FF_0 = (v_{oc} - \ln(v_{oc} + 0:71))/(v_{oc} + 1), \tag{5}$$

where v_{oc} is

$$v_{oc} = V_{oc}/(nkT/q). \tag{6}$$

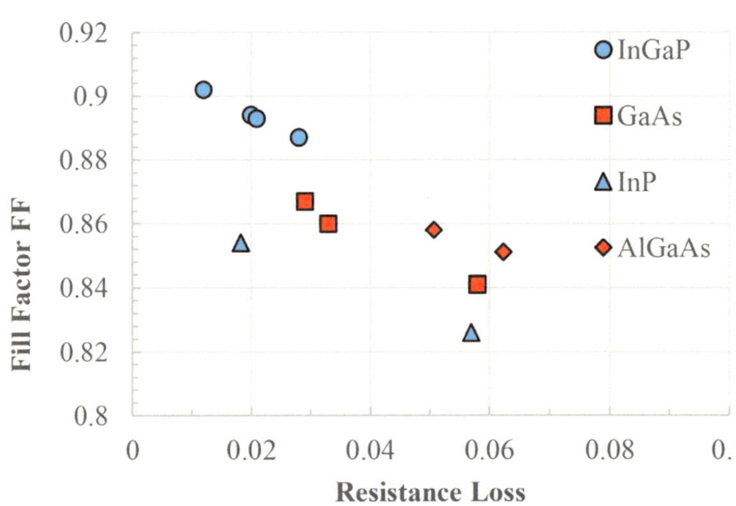

Figure 4.
Correlation between fill factor and resistance loss in GaAs, InP, AlGaAs and InGaP solar cells.

The measured fill factors can then be related to the series resistance and shunt resistance by the following Eq. [18]:

$$FF \approx FF_0(1 - r_s)(1 - r_{sh}^{-1}) \approx FF_0(1 - r_s - r_{sh}^{-1}) = FF_0(1 - r), \tag{7}$$

where r_s is the series resistance, and r_{sh} is the shunt resistance normalized to R_{CH}. The characteristic resistance R_{CH} is defined by [18]

$$R_{CH} = V_{oc}/J_{sc}, \tag{8}$$

r is the total normalized resistance defined by $r = r_s + r_{sh}^{-1}$.

Figure 4 shows correlation between fill factor and resistance loss [2, 8–13, 17] in GaAs, InP, AlGaAs and InGaP solar cells. Lower resistance losses of 0.01-0.03 have been realized for GaAs, InP and InGaP solar cells compared to 0.05 for AlGaAs.

3. Historical progress and key issues for high-efficiency III-V compound single-junction solar cells

Table 2 shows major losses, their origins and key technologies for improving efficiency [6]. There are several loss mechanisms to be solved for realizing high-efficiency III-V compound single-junction solar cells. (1) bulk recombination loss, (2) surface recombination loss, (3) interface recombination loss, (4) voltage loss, (5) fill factor loss, (6) optical loss, (7) insufficient –energy photon loss. Key

Losses	Origins	Technologies for improving
Bulk recombination loss	Non radiative recombination centers (impurities, dislocations, grain boundary, other defects)	High quality epitaxial growth Reduction in density of defects
Surface recombination loss	Surface sates	Surface passivation Heteroface layer Double hetero structure
Interface Recombination loss	Interface states Lattice mismatching defects	Lattice matching Inverted epitaxial growth Window layer Back surface field layer Double hetero structure Graded band-gap layer
Voltage loss	Non radiative recombination Shunt resistance	Reduction in density of defects Thin layer
Fill factor loss	Series resistance Shunt resistance	Reduction in contact resistance Reduction in leakage current, Surface, interface passivation
Optical loss	Reflection loss Insufficient absorption	Anti-reflection coating, texture Back reflector, photon recycling
Insufficient-energy photon loss	Spectral mismatching	Multi-junction (Tandem) Down conversion Up conversion

Table 2.
Major losses, their origins of III-V compound cells and key technologies for improving efficiency.

technologies for reducing the above losses are high quality epitaxial growth, reduction in density of defects, optimization of carrier concentration in base and emitter layers, double-hetero (DH) structure junction, lattice-matching of active layers and substrate, surface and interface passivation, reduction in series resistance and leakage current, anti-reflection coating, photon recycling and so forth.

Solar cell efficiency is dependent upon minority-carrier diffusion length (or minority-carrier lifetime) in the solar cell materials as shown in **Figure 5**.

Radiative recombination lifetime τ_{rad} is expressed by

$$\tau_{rad} = 1/BN \tag{9}$$

where N is the carrier concentration and B is the radiative recombination probability. The B value for GaAs reported by Ahrenkiel et al. [19] is B = 2 X 10^{-10} cm^3/s. Effective lifetime τ_{eff} is expressed by

$$1/\tau_{eff} = 1/\tau_{rad} + 1/\tau_{nonad} \tag{10}$$

where τ_{nonrad} is non-radiative recombination lifetime and given by

$$1/\tau_{nonrad} = \sigma v N_r \tag{11}$$

where σ is capture cross section of minority-carriers by non-radiative recombination centers, v is minority-carrier thermal velocity, and Nr is density of non-radiative recombination center.

Therefore, improvement in crystalline quality and reduction in densities of defects such as dislocations, grain boundaries and impurities that act as non-radiative recombination centers are very important for realizing high-efficiency solar cells.

In this chapter, analytical results for historical progress in efficiency of GaAs single-junction solar cells are shown. **Figures 6** and 7 show analytical results for progress in ERE and resistance loss of GaAs single-junction solar cells.

The first GaAs solar cells reported by Jenny et al. [20] were fabricated by Cd diffusion into an n-type GaAs single crystal wafer. Efficiencies of 3.2-5.3% were quite low due to deep junction. Because GaAs has large surface recombination velocity S of

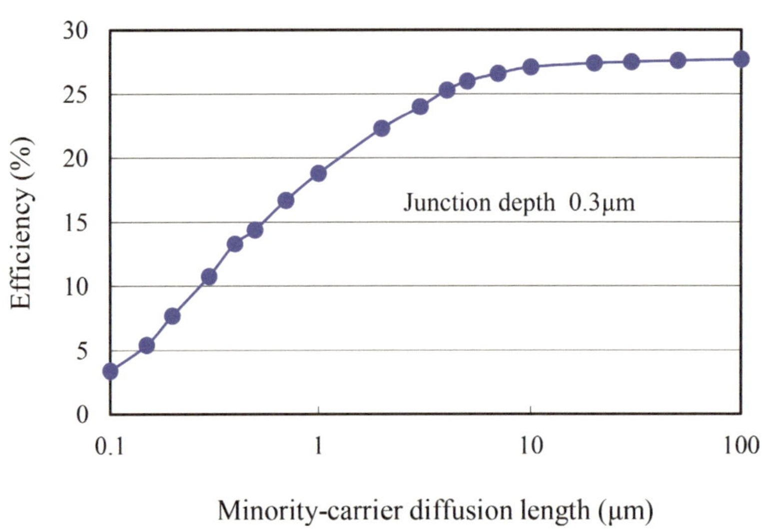

Figure 5.
Minority-carrier diffusion length dependence of GaAs solar cell characteristics.

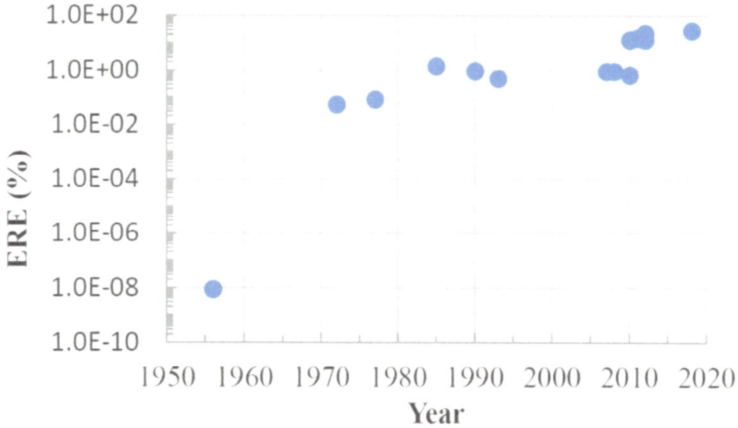

Figure 6.
Analytical results for ERE progress of GaAs single-junction solar cells.

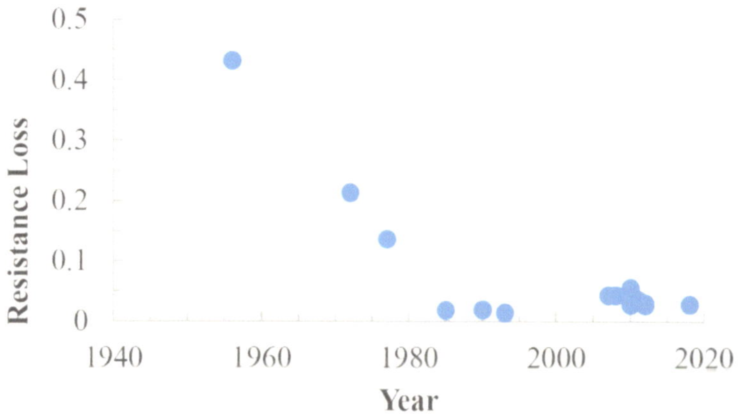

Figure 7.
Analytical results for resistance loss progress of GaAs single-junction solar cells.

around 1×10^7 cm/s [6, 21], formation of shallow homo-junction with junction depth of less than 50 nm is necessary to obtain high-efficiency. Therefore, hetero-face structure AlGaAs-GaAs solar cells have been proposed by Woodall and Hovel [22] and more than 20% efficiency has been realized [22] in 1972 as shown in **Figure 1** as a result of ERE improvement from 10^{-8}% to 0.05% as shown in **Figure 6**. Double-hetero (DH) structure AlGaAs-GaAs-AlGaAs solar cell with an efficiency of 23% has been realized by Fan's group in 1985 [23] as a result of ERE improvement from 0.05% to 1.4% as shown in **Figure 6**. Now, DH structure solar cells are widely used for high-efficiency III-V compound solar cells including GaAs solar cells.

Figure 8 shows device structures of GaAs solar cells developed historically. As mentioned above and shown in **Figure 8**, device structures of GaAs cells were improved from homo-junction, to heteroface structure, finally to DH structure. Now, InGaP layer is mainly used as front window and rear back surface field (BSF) layers instead of AlGaAs layer. The reasons are explained in the part of multi-junction solar cells.

Figure 9 shows the chronological improvements in the efficiencies of GaAs solar cells fabricated by LPE (Liquid Phase Epitaxy), MOCVD (Metal-Organic Chemical Vapor Deposition) and MBE (Molecular Beam Epitaxy). LPE was used to fabricate AlGaAs-GaAs heteroface solar cells in 1972 because it produces high-quality

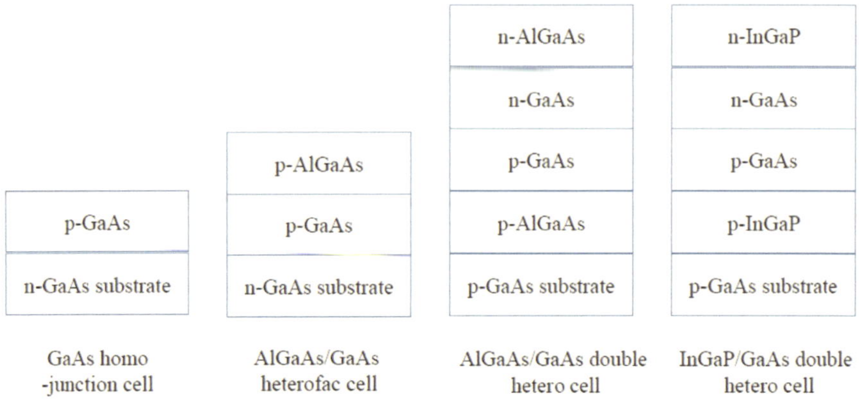

Figure 8.
Device structures of GaAs solar cells developed historically.

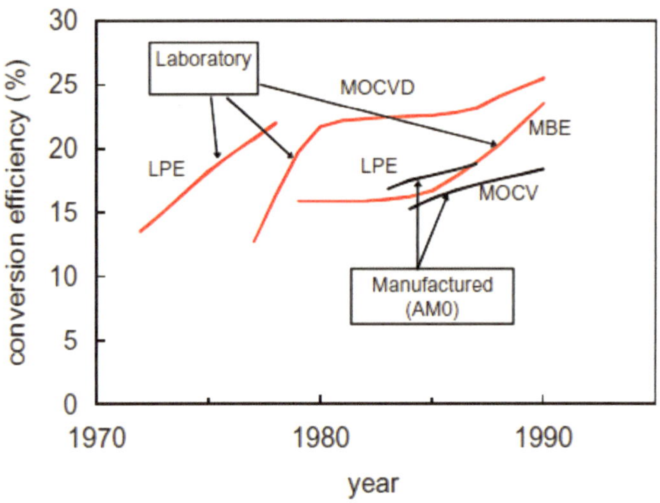

Figure 9.
Chronological improvements in the efficiencies of GaAs solar cells fabricated by the LPE, MOCVD and MBE methods.

epitaxial film and has a simple growth system. Homo-junction structure and heteroface structure GaAs solar cells shown in **Figure 8** were fabricated by LPE. However, it is not as useful for devices that involve multilayers because of the difficulty of controlling layer thickness, doping, composition and speed of throughput. Since 1977, MOCVD has been used to fabricate large-area GaAs solar cells by using DH structure shown in **Figure 8** because it is capable of large-scale, large-area production and has good reproducibility and controllability.

Regarding the differences of surface recombination velocities in semiconductor materials, differences of point defect behavior are thought to be one of the mechanisms. For example, because nearest-neighbor hopping migration energies (0.3 eV and 1.2 eV) of V_{In} and V_P in InP [24] are lower than those (1.75 eV) of V_{Ga} and V_{As} in GaAs, better surface state may be formed on InP surface compared to GaAs surface.

In addition to improvement in surface recombination loss, as a result of technological development, resistance loss has been improved as shown in **Figure** 7. In parallel, bulk recombination loss and interface recombination loss have been

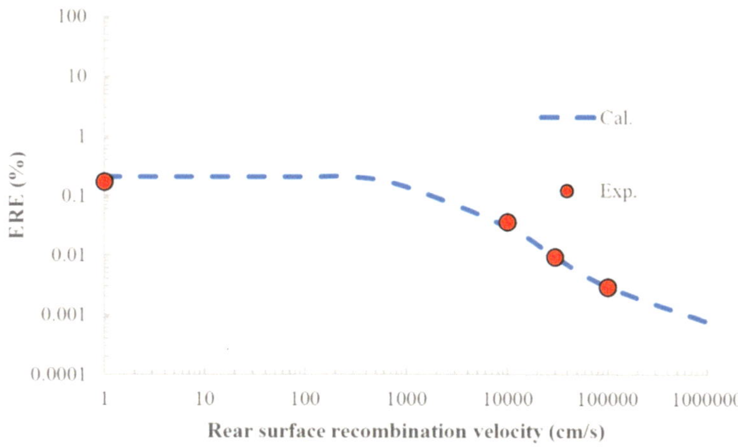

Figure 10.
Correlation between ERE and interface recombination velocity in InGaP single-junction solar cells.

improved as shown in **Figure 6**. Recently, efficiency of GaAs solar cells reached to 29.1% [2] by realizing ERE of 22.5% as a result of effective photon recycling [1].

Lattice mismatching also degrades solar cell properties by increase in interface recombination velocity as a result of misfit dislocations and threading dislocations generation. By using interface recombination velocity S_I as a function of lattice mismatch ($\Delta a/a_0$) for InGaP/GaAs heteroepitaxial interface [25], lattice mismatch ($\Delta a/a_0$) dependence of interface recombination velocity (S_I) is semi-empirically expressed by [16].

$$S_I\ [\mathrm{cm/s}] = 1.5 \times 10^8 \Delta a/a_0 \qquad (12)$$

As one of example for effects of interface recombination loss upon solar cell properties, analytical results for correlation between ERE and interface recombination velocity in InGaP single-junction solar cells are shown in **Figure 10**.

4. Historical progress and key issues for high-efficiency III-V compound multi-junction solar cells

While single-junction cells may be capable of attaining AM1.5 efficiencies of up to 30-32% as shown in **Figure 2**, the multi-junction (MJ) structures [26, 27] were recognized early on as being capable of realizing efficiencies of up to 46% as shown below. **Figure 11** shows the principle of wide photo response using MJ solar cells for the case of a triple-junction cell. Solar cells with different bandgaps are stacked one on top of the other so that the cell facing the Sun has the largest bandgap (in this example, this is the InGaP top cell). This top cell absorbs all the photons at and above its bandgap energy and transmits the less energetic photons to the cells below. The next cell in the stack (here the GaAs middle cell) absorbs all the transmitted photons with energies equal to or greater than its bandgap energy, and transmits the rest downward in the stack (in this example, to the Ge bottom cell). As shown in **Figure 12**, the spectral response for an InGaP/GaAs/Ge monolithic, two-terminal triple-junction cell shows the wideband photo response of multijunction cells. In principle, any number of cells can be used in tandem.

As a result of research and development, high-efficiencies have been demonstrated with III-V multi-junction solar cells: 37.9% under 1-sun and 44.4% under

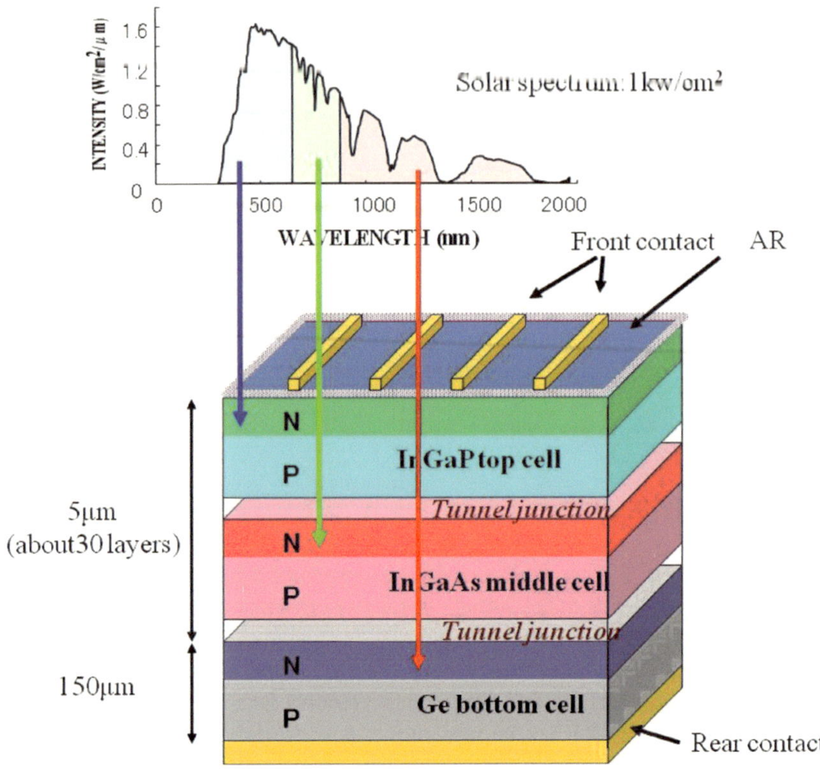

Figure 11.
Principle of wide photo response by using a multijunction solar cell, for the case of an InGaP/GaAs/Ge triple-junction solar cell.

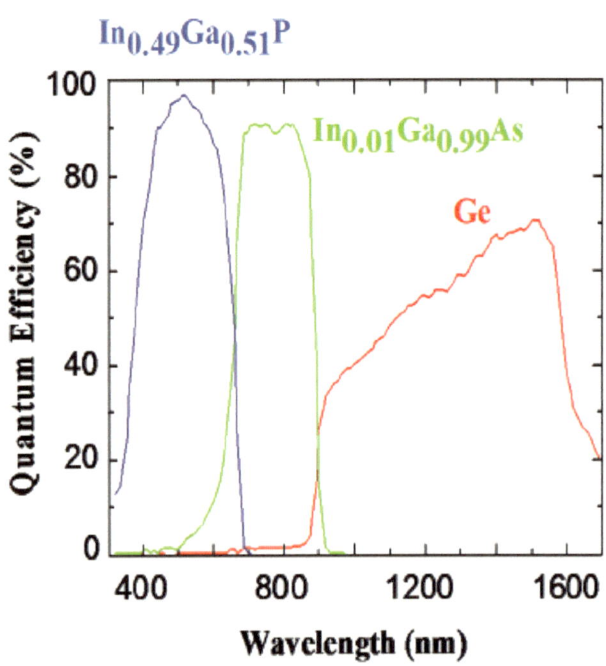

Figure 12.
Spectral response for an InGaP/GaAs/Ge monolithic, two-terminal three-junction cell.

concentration for 3-junction cells [28] and 39.2% under 1-sun, 47.1% under concentration for 6-junction solar cells [7]. **Figure 13** shows historical record-efficiency of III-V multi-junction (MJ) and concentrator MJ solar cells in comparison with 1-sun efficiencies of GaAs and crystalline Si solar cells, along with their extrapolations [3].

Table 3 shows key issues for realizing super high-efficiency MJ solar cells. The key issues for realizing super-high-efficiency MJ solar cells are (1) sub cell material selection, (2) tunnel junction for sub cell interconnection, (3) lattice-matching, (4) carrier confinement, (5) photon confinement, (6) anti-reflection in wide wavelength region and so forth. For concentrator applications by using MJ cells, the cell front contact grid structure should be designed in order to reduce the energy loss due to series resistance (resistances of front grid electrode including contact resistance, rear electrode, lateral resistance between grid electrodes) by considering shadowing loss attributed to grid electrode, and tunnel junction with high tunnel peak current density is necessary. Because cell interconnection of sub-cells is one of the most important key issues for realizing high-efficiency MJ solar cells in order to reduce losses of electrical connection and optical absorption, effectiveness of double hetero structure tunnel diode is also presented in this chapter.

Selection of sub-cell layers by considering optimal bandgap and lattice matching of materials is one of key issues for realizing super high-efficiency MJ cells. **Table 4** shows one example for selection of top cell material and comparison of InGaP and AlGaAs as a top cell material. InGaP that has better interface recombination velocity, less oxygen-related defect problems and better window material AlInP compared to those of AlGaAs has been proposed as a top cell material by NREL group [29]. As described above, InGaP materials are now widely used as front widow and back surface filed layers of solar cells instead of AlGaAs.

Figure 14 shows the connection options for two-junction cells: the two cells can be connected to form either two-terminal, three-terminal or four-terminal devices. In a monolithic, two-terminal device, the cells are connected in series with an optically transparent tunnel junction intercell electrical connection. In a two-terminal structure, only one external circuit load is needed, but the photocurrents in

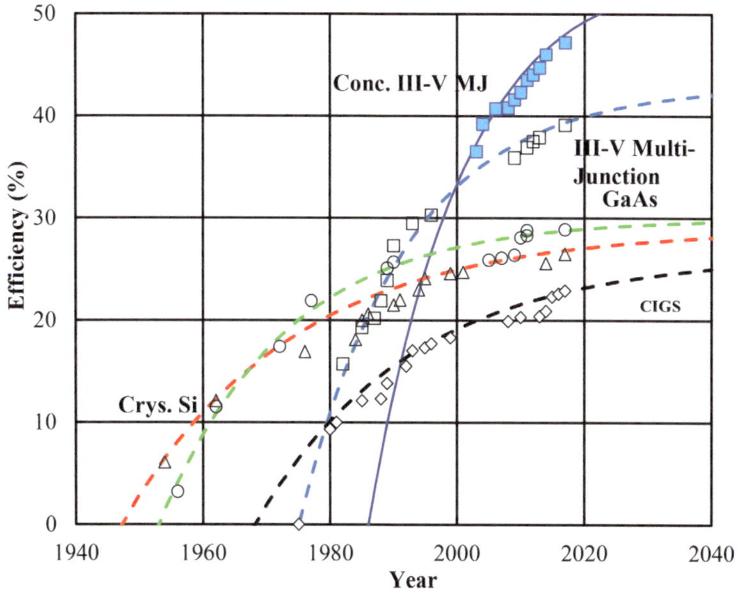

Figure 13.
Historical record-efficiency of III-V multi-junction (MJ) and concentrator MJ solar cells in comparison with 1-sun efficiencies of GaAs and crystalline Si solar cells, along with their extrapolations.

Key issue	Past	Present	Future
Top cell materials	AlGaAs	InGaP	AlInGaP
Middle cell materials	None	GaAs, InGaAs	GaAs, quantum well, quantum dots, InGaAs, InGaAsN etc.
Bottom cell materials	GaAs	Ge, InGaAs	Si, Ge, InGaAs
Substrate	GaAs	Ge	Si, Ge, GaAs, metal
Tunnel junction (TJ)	Double hetero structure-GaAs TJ	Double hetero structure-InGaP TJ	Double hetero structure-InGaP or GaAs TJ
Lattice matching	GaAs middle cell	InGaAs middle cell	(In)GaAs middle cell
Carrier confinement	InGaP-BSF	AlInP-BSF	Wide-gap-BSF Quantum dots
Photon confinement	None	None	Back reflector, Bragg reflector, quantum dots, photonic crystals, etc.
Others	None	Inverted epitaxial growth	Inverted epitaxial growth, epitaxial lift off

Table 3.
Key issues for realizing super-high-efficiency III-V compound multi-junction solar cells.

	InGaP	AlGaAs
Interface recombination velocity	$<5 \times 10^3$ cm/s	10^4–10^5 cm/s
Oxygen-related defects	Less	Higher
Window Layer (Eg)	AlInP (2.5 eV)	AlGaAs (2.1 eV)
Other problems	High doping in p-AlInP	Lower efficiency (2.6% lower)

Table 4.
Comparison of InGaP and AlGaAs as a top cell material.

the two cells must be equal for optimal operation. Key issues for maximum-efficiency monolithic cascade cells (two-terminal multijunction cells series connected with tunnel junction XE "tunnel junctions") are the formation of tunnel junctions of high performance and stability for cell interconnection, and the growth of optimum bandgap top- and bottom-cell structures on lattice-mismatched substrates, without permitting propagation of deleterious misfit and thermal stress-induced dislocations.

As shown in **Table 3**, cell interconnection of sub-cells is one of the most important key issues for realizing high-efficiency MJ solar cells. DH structure has been found to effectively prevent from impurity diffusion from tunnel junction and high tunnel peak current density has been obtained by the authors [30, 31]. **Figure 15** shows annealing temperature (equivalent to growth temperature of top cell layers) dependence of tunnel peak current densities for double hetero structure tunnel diodes. X is the Al mole fraction in $Al_xGa_{1-x}As$ barrier layers [30, 31]. It has also been found that the impurity diffusion from the tunnel junction is effectively suppressed by the wider bandgap material tunnel junction with wider bandgap material-double hetero (DH) structure [32]. These results are thought to be due to the lower diffusion coefficient for impurities in the wider band gap materials such as the AlInP barrier layer and InGaP tunnel junction layer [32].

As a result of developing high performance tunnel junction with high tunnel peak current density, high efficiency MJ solar cells have been developed [30, 33, 34].

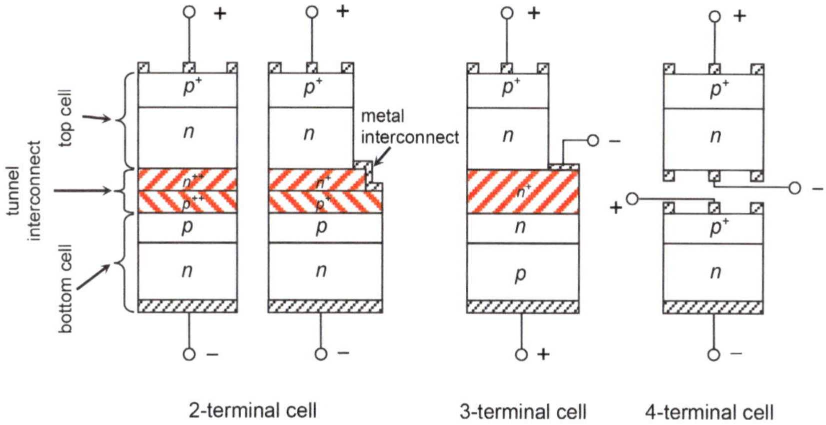

Figure 14.
Schematic diagrams of various configurations of two-junction cells.

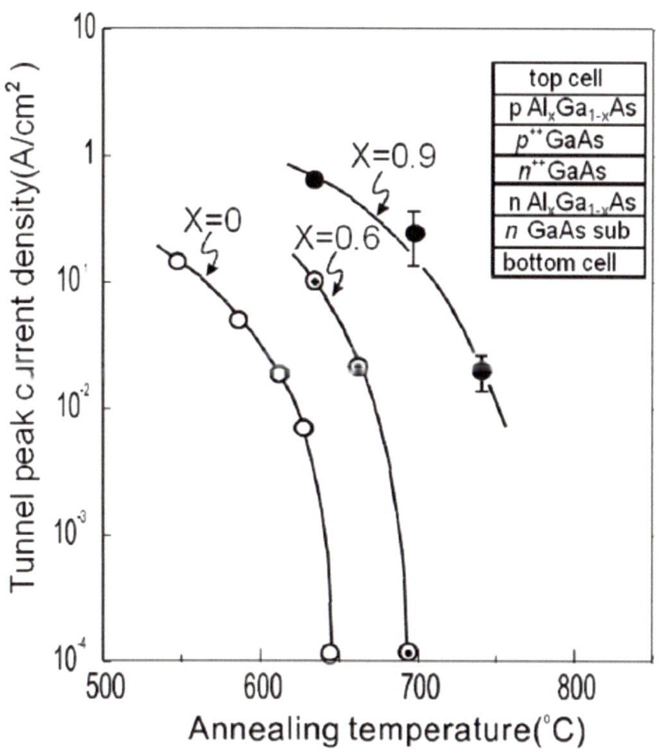

Figure 15.
Annealing (growth) temperature dependence of tunnel peak current densities for double hetero structure tunnel diodes. X is the Al mole fraction in $Al_xGa_{1-x}As$ barrier layers.

Figure 16 shows a structure and light-illuminated (AM1.5G 1-sun) I-V characteristics of InGaP/GaAs/InGaAs 3-junctuon solar cell. 37.9% efficiency under AM1.G 1-sun and 44.4% under 300-suns concentration have been demonstrated with InGaP/GaAs/InGaAs 3-junction solar cell by Sharp [35]. Spectrolab has achieved 38.8% efficiency under 1-sun with 5-junction solar cells [36]. FhG-ISE has demonstrated 46.0% under 58-suns concentration with 4-junction solar cells [37]. Most recently, 39.2% under AM1.5 1-sun and 47.1% under 144-suns have been realized with 6-junction cell by NREL [7].

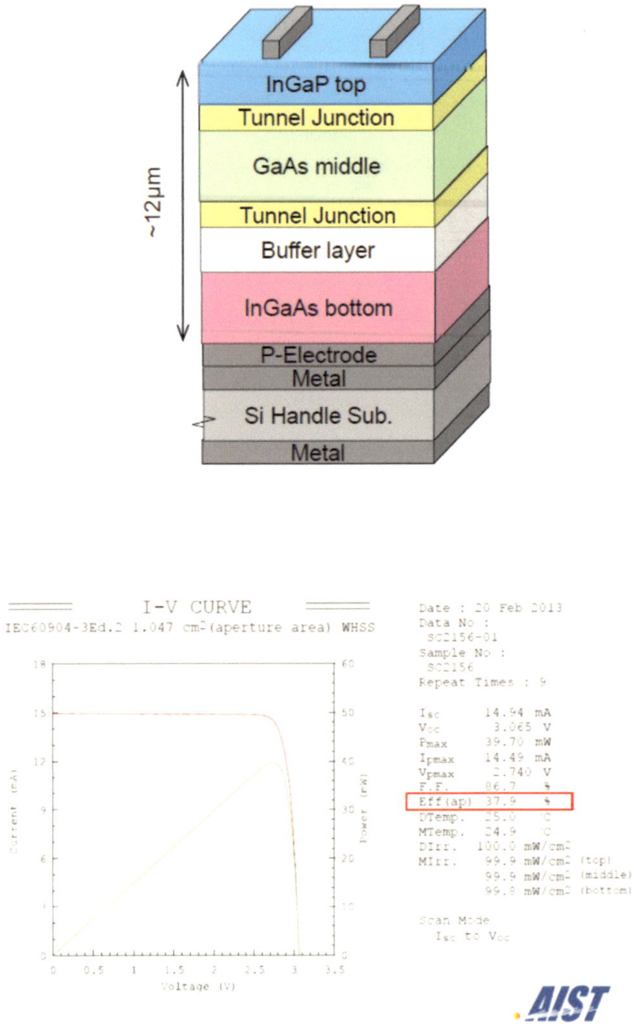

Figure 16.
A structure and light-illuminated I-V characteristics of InGaP/GaAs/InGaAs 3-junctuon solar cell.

5. Radiation resistance and space applications of III-V compound single-junction and multi-junction solar cells

Development radiation-resistant solar cells is necessary for space application because solar cells degrade due to defect generation under radiation environment in space. Recombination centers tend to affect the solar cell performance by reducing the minority carrier diffusion length L in solar cell active layer from a pre-irradiation value L_0 to a post-irradiation value L_φ through Eq.

$$1/{L_\varphi}^2 - 1/{L_0}^2 = \Sigma I_{ri}\sigma_i v_{th}\varphi/D = K_L\varphi, \qquad (13)$$

where suffixes 0 and φ show before and after irradiation, respectively, I_{ri} is introduction rate of i-th recombination center by electron irradiation, σ_i the capture cross section of minority-carrier by i-th recombination center, v_{th} the thermal velocity of minority-carrier, D the minority-carrier diffusion coefficient, K_L the damage coefficient for minority-carrier diffusion length, and φ the electron fluence.

The III-V compound solar cells have better radiation tolerance compared to crystalline Si cells because many III-V compound materials have direct band gap and higher optical absorption coefficient compared to Si with in-direct bandgap. In addition, InP-related materials such as InP, InGaP, AlInGaP, InGaAsP are superior

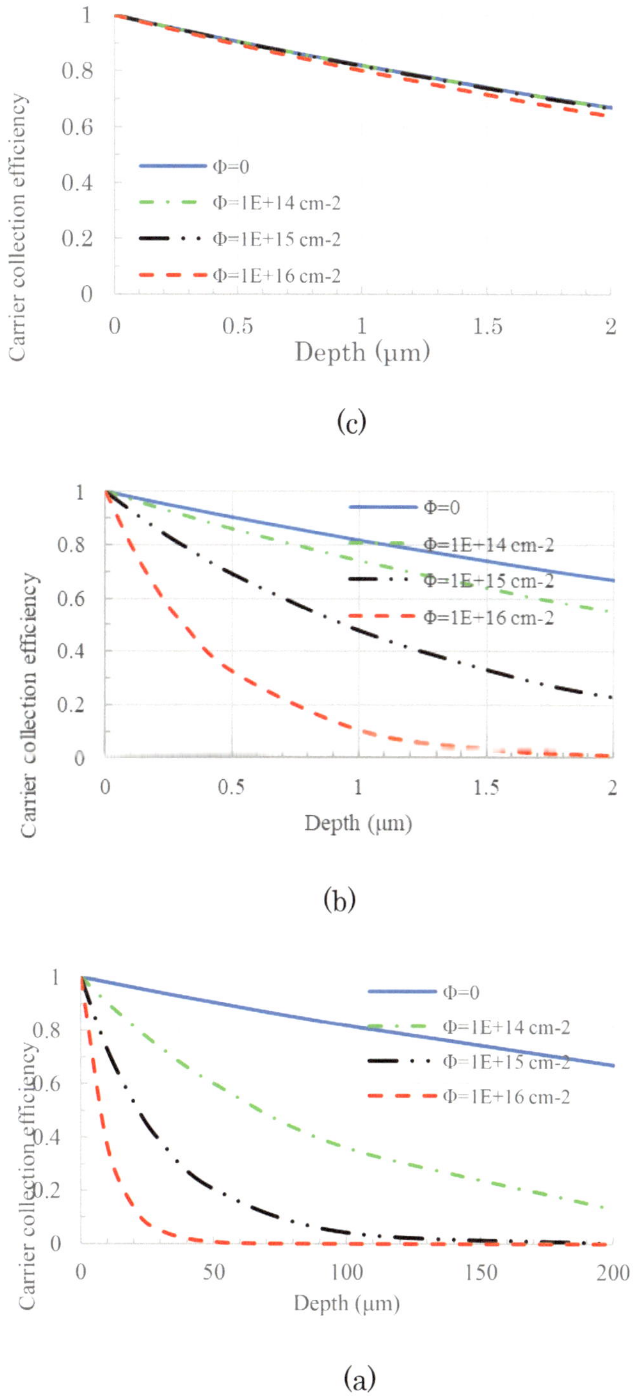

Figure 17.
Calculated depth × distribution of carrier collection efficiency in (a) Si, (b) GaAs and (c) InP under 1-MeV electron irradiation, calculated by using our experimental values [40–42] and Eq. (11), and by assuming carrier collection efficiency as a function of exp.(−x/L).

radiation-resistant compared to Si and GaAs and have unique properties that radiation-induced defects in InP-related materials are annihilated under minority-carrier injection such as light-illumination at room temperature or low temperature of less than 100 K [38, 39].

Figure 17 shows calculated depth x distribution of carrier collection efficiency in Si, GaAs and InP under 1-MeV electron irradiation, calculated by using our experimental values [40–42] and Eq. (13), and by assuming carrier collection efficiency as a function of exp.(−x/L). It is clear from **Figure 17** that GaAs has better radiation-tolerance and InP has superior radiation tolerance compared to Si.

Figure 18 shows changes in efficiency of Si single-junction, GaAs single-junction and InGaP/GaAs/Ge 3-junction space solar cells as a function of 1-MeV electron

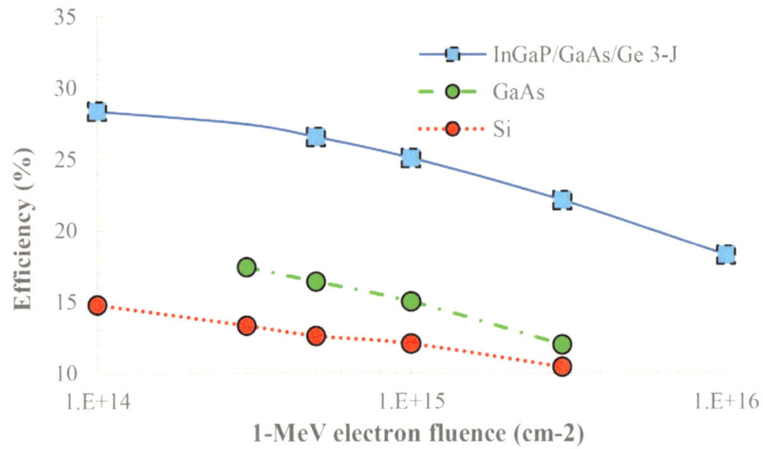

Figure 18.
Changes in efficiency of Si single-junction, GaAs single-junction and InGaP/GaAs/Ge 3-junction space solar cells as a function of 1-MeV electron fluence.

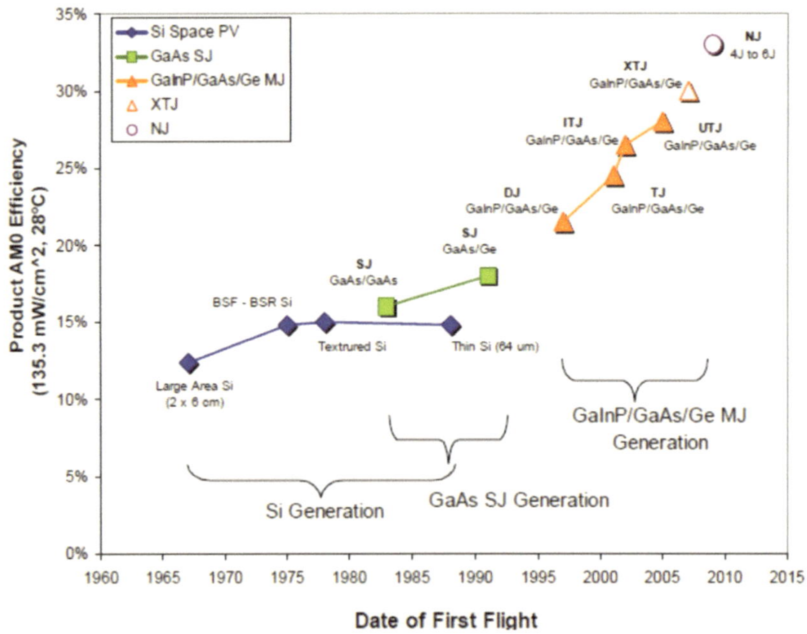

Figure 19.
Historical product efficiency of space solar cells against date of first flight. Open points are for planned products and estimate flight dates.

fluence. The InGaP/GaAs/Ge 3-junction solar cells is now mainly used for space as shown below because they are radiation-resistant and are highly efficient compared to Si and GaAs space solar cells [43].

Because GaAs single-junction solar cells and III-V compound multi-junction solar cells have high-efficiency and radiation-resistance compared to Si solar cells, III-V compound solar cells are mainly used in space as shown in **Figure 19** [44].

6. Future prospects

The multijunction solar cells will be widely used in space because of their high conversion efficiency and good radiation resistance. However, in order to apply super-high-efficiency cells widely on Earth, it will be necessary to improve their conversion efficiency and reduce their cost. **Figure 20** summarizes efficiency potential of single-junction and multi-junction solar cells, calculated by using the similar procedure presented in Section 2, in comparison with experimentally realized efficiencies under 1sun illumination. Altough single-junction solar cells have potential efficiencies of less than 32%, 3-junction and 6-junction solar cells have potential efficiencies of 42% and 46%, respectively.

The concentrator PV (CPV) systems [45] with several times more annual power generation capability than conventional crystalline silicon flat-plate systems will open a new market for apartment or building rooftop and charging stations for battery powered electric vehicle applications. Other interesting applications are in agriculture and large-scale PV power plants.

The multi-junction solar cells are greatly expected as high-efficiency solar cells into solar cell powered electric vehicles. **Figure 21** shows required conversion efficiency of solar modules as a function of its surface area and electric mileage to attain 30 km/day driving. A preferable part of the installation is the vehicle roof. Because of space limitation for passenger cars, development high-efficiency solar cell modules with efficiencies of more than 30% is very important as shown in **Figure 21** [46, 47].

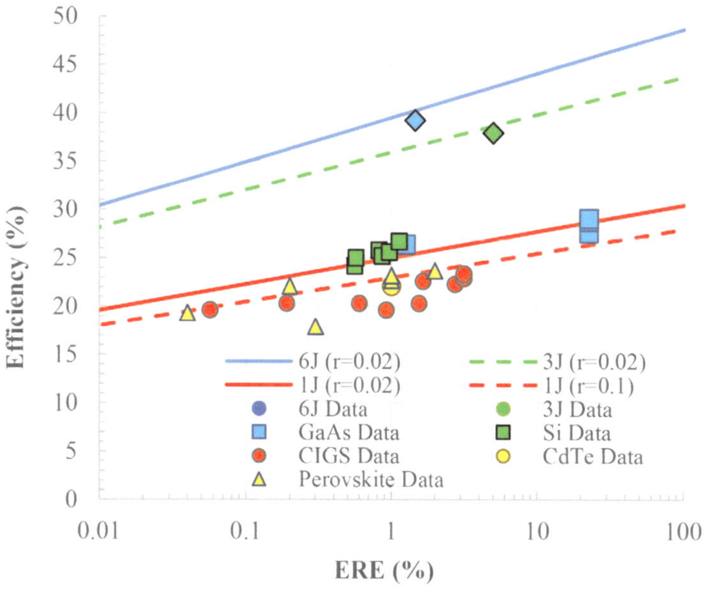

Figure 20.
Calculated conversion efficiencies of various single-junction, 3-junction and 6-junction solar cells, calculated by using the similar procedure presented in Section 2, in comparison with experimentally realized efficiencies under 1-sun illumination.

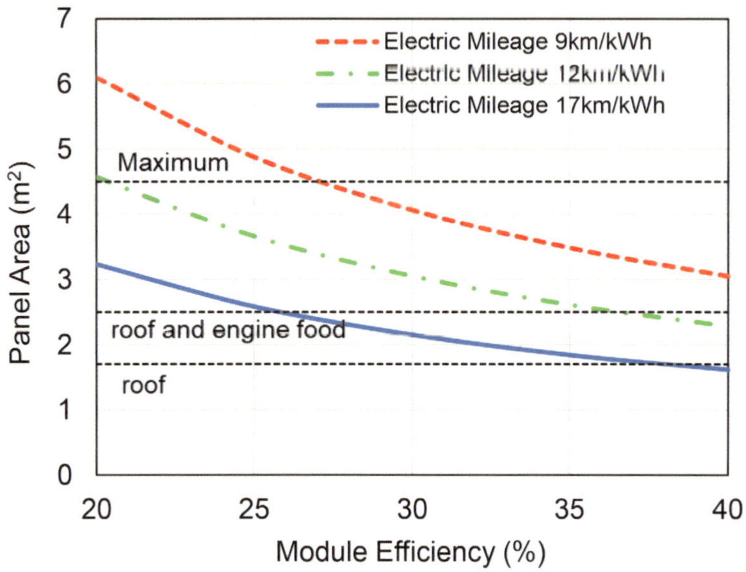

Figure 21.
Required conversion efficiency of solar modules as a function of its surface area and electric mileage to attain 30 km/day driving. A preferable part of the installation is the vehicle roof.

In addition to high-efficiency, cost reduction of solar cell modules is necessary. Therefore, further development of high-efficiency and low-cost modules is necessary.

7. Conclusion

This chapter reviewed progress in GaAs-based single junction solar cells and III-V compound multi-junction solar cells and key issues for realizing high-efficiency solar cells. The III-V compound solar cells have contributed as space and concentrator solar cells and are expected as creation of new markets such as large-scale electric power systems and solar cell powered electric vehicles. Regarding single-junction solar cells, especially, GaAs solar cells have shown 29.1% under 1-sun illumination, highest ever reported for single-junction solar cells. In addition, analytical results for non-radiative recombination and resistance losses in III-V compound solar cells are shown by considering fundamentals for major losses in III-V compound materials and solar cells. Because the limiting efficiency of single-junction solar cells is 30-32%, multi-junction junction solar cells have been developed and InGaP/GaAs based 3-junction solar cells are widely used in space. The InGaP/GaAs/InGaAs 3-junction solar cells have recorded 37.4% under 1-sun and 44.4% under concentration. Recently, highest efficiencies of 39.1% under 1-sun and 47.2% under 144-suns concentration have been demonstrated with 6-junction solar cells. The 3-junction and 6-junction solar cells potential efficiencies of 42% and 46% under 1-sun, respectively. Further development of high-efficiency and low cost solar cells and modules is necessary in order to create new markets.

Acknowledgements

Our studies were partially supported by the NEDO (New Energy and Industrial Technology Development Organization) and JSPS (Japan Society for Promotion of

Science). The author wishes to express sincere thanks to Dr. T. Takamoto, Sharp, Dr. K. Araki, Toyota Tech. Inst., Dr. M. Imaizumi, JAXA, Dr. A. Yamamoto, Fukui Univ., Dr. H. Sugiura and Dr. C. Amano, formerly NTT Lbs., Dr. SJ. Taylor, ESA, Prof. A. Kahn, South Arabama Univ., Prof. HS. Lee, Korea Univ., Prof. N. Ekins-Daukes, UNSW, Prof. A. Luque, UPM, Dr. A. Bett Dr. G. Sifer and Dr. F. Dimroth, FhG-ISE, Dr. M. Al-Jassim, Dr. R. Ahrenkiel and Dr. J.F. Geisz, NREL for their fruitful collaboration and discussion.

Conflict of interest

The author declares no conflict of interest.

Author details

Masafumi Yamaguchi
Toyota Technological Institute, Nagoya, Japan

*Address all correspondence to: masafumi@toyota-ti.ac.jp

References

[1] Kayes BM, Nie H, Twist R, Spruytte SG, Reinhardt F, Kizilyalli IC, Higashi GS. 27.6% conversion efficiency, a new record for single junction solar cells under 1 sun illumination. Proceedings of the 37th IEEE Photovoltaic Specialists Conference, IEEE;2011:4. DOI: 10.1109/PVSC.2011.6185831

[2] Green MA, Dunlop ED, Hohl-Ebinger J, Yoshita M, Kopidakis N, Hao X. Solar cell efficiency tables (Version 56). Progr. Photovoltaics. 2020; 28: 629. DOI: 10.1002/pip.3303.

[3] Yamaguchi M, Lee KH, Araki K, Kojima N. A review of recent progress in heterogeneous silicon tandem solar cells. J. Phys. D. Appl. Phys. 2018;51: 133002. DOI: 10.1088/1361-6463/aaaf08. and updated.

[4] Goetzberger A, Luther J, Willek G. Solar cells: past, present, future. Solar Energy Mater. Solar Cells. 2002;74:1. DOI: 10.1016/S0927-0248(02)00042-9

[5] Shockley W, Queisser HJ. Detailed balance limit of efficiency of pn junction solar cells. J. Appl. Phys. 1961;32:510. DOI: 10.1063/1.1736034

[6] Yamaguchi M. Fundamentals and R&D status of III-V compound solar cells and materialsPhys. Satus Solidi. 2015;C12:489. DOI: 10.1002/pssc.201400216

[7] Geisz JF, France RM, Schulte KL, Steiner MA, Norman AG, Guthrey HL, Young MR, Song T, Moriarty T. Six-junction III-V solar cells with 47.1% conversion efficiency under 143 suns concentration. Nature Energy, 2020;5: 326. DOI: 10.1038/s41560-020-0598-5.

[8] Yamaguchi M, Yamada H, Katsumata Y, Lee KH, Araki K, Kojima N. Efficiency potential and recent activities of high-efficiency solar cells. J. Mater. Res. 2017;32:3445. DOI: 10.1557/jmr.2017.335

[9] Yamaguchi M, Lee KH, Araki K, Kojima N, Yamada H, Katsumata Y. Analysis for efficiency potential of high-efficiency and next generation solar cells. Prog Photovol. 2018;26:543. DOI: 10.1002/pip.2955

[10] Green MA, Emery K, Hishikawa Y, Warta W. Solar cell efficiency tables (version 36). Prog. Photovoltaics. 2010; 18:346. DOI: 10.1002/pip.1021

[11] Green MA, Emery K, Hishikawa Y, Warta W, Dunlop ED. Solar cell efficiency tables (version 45). Prog. Photovoltaics. 2016;24:905. DOI: 10.1002/pip.2788

[12] Green MA, Emery K, Hishikawa Y, Warta W, Dunlop ED, Levi DH, Ho-Baillie AWY. Prog. Solar cell efficiency tables (version 50). Prog. Photovoltaics. 2017;25:668. DOI: 10.1002/pip.2909

[13] Green MA, Dunlop ED, Levi DH, Hohl-Ebinger J, Yoshita M, Ho-Baillie AWY. Solar cell efficiency tables (version 54). Prog. Photovoltaics. 2019; 27,565. DOI: 10.1002/pip.3171.

[14] Rau U. Reciprocity relation between photovoltaic quantum efficiency and electroluminescent emission of solar cells. Phys. Rev. 2007;B76:085303. DOI: 10.1103.

[15] Yao J, Kirchartz T, Vezie MS, Faist MA, Gong W, He Z, H. Wu H, Troughion J, Watson T, Bryant D, Nelson J. Quantifying losses in open-circuit voltage in solution-processable solar cells. Phys. Rev. Applied. 2015;4: 014020. DOI: 10.1103/PhysRevApplied.4.014020

[16] Green MA. Radiative efficiency of state-of-the-art photovoltaic cells. Prog.

Photovoltaics. 2012;20:472. DOI: 10.1002/pip.1147

[17] Geisz JF, Steiner MA, Garcia I, Kurtz SR, Fiedman DJ. Enhanced external radiative efficiency for 20.8% efficient single-junction GaInP solar cells. Appl. Phys. Lett. 2013;103:041118. . DOI: 10.1063/1.4816837.

[18] Green MA. Solar Cells. UNSW, Kensington; 1998. ISBN-10: 0138222703.

[19] Ahrenkiel R. Minority-carrier lifetime in III-V semiconductors. Semiconductors and Semimetals, Vol. 39, eds. Ahrenkiel RK and Lundstrom MS. Academic Press; 1993: 58. DOI: 1016/S0080-8784(08)62594-6

[20] Jenny DA, Loferski JJ, Rappaport P. Photovoltaic effect in GaAs p-n junctions and solar energy conversion. Phys. Rev. 1956;101:1208. DOI: 10.1103/ PhysRev.101.1208.

[21] Casey HC,Jr., Beehler E. Evidence for low surface recombination velocity on n type InP. Appl. Phys. Lett. 1977,30. 247. DOI: 10.1063/1.89352.

[22] Woodal JM, Hovel HJ. High-efficiency $Ga_{1-x}Al_xAs$–GaAs solar cells. Appl. Phys. Lett. 1972;21:379. DOI: 10.1063/1.1654421

[23] Gale RP, Fan JCC, Turner GW, Chapman RL. High-efficiency double-heterostructure AlGaAs/GaAs solar cells. Proceedings of the 18th IEEE Photovoltaic Specialists Conference. IEEE;1985:296.

[24] Van Vechten JA, Wagner JF. Asymmetry of anion and cation vacancy migration enthalpies in III-V compound semiconductors: Role of the kinetic energy. Phys. Rev. 1985;B32:5259. DOI: 10.1103/PhysRevB.32.5259.

[25] Neuse CJ. III-V alloys for optoelectronic applications. J. Electron. Mater. 1977;6:253. DOI: 10.1007/ BF02660488.

[26] Yamaguchi M. Super-high efficiency III-V tandem and multijunction cells. Eds. Archer M and Green MA. Clean Electricity from Photovoltaics. World Scientific; 2001: 347. DOI: 10.1142/9781848161504_ 0008.

[27] Yamaguchi M, Takamoto T, Araki K, Ekins-Daukes N. Multi-junction III-V solar cells: current status and future potential. Solar Energy. 2005;79:78. DOI: 10.1016/j.solene r.2004.09.018.

[28] Yamaguchi, Takamoto T, Araki K, Kojima N. Recent results for concentrator photovoltaics in Japan. Jpn. J. Appl. Phys. 2016;55:04EA05. DOI: 10.7567/JJAP.55.04EA05.

[29] Olson JM, Kurtz SR, Kibbler AE. A 27.3% efficient $Ga_{0.5}In_{0.5}P$/ GaAs tandem solar cell. Appl. Phys. Lett. 1990;56:623. DOI: 10.1063/1.102717.

[30] Yamaguchi M., Amano C., Sugiura H. Yamamoto A. High efficiency AlGaAs/GaAs tandem solar cells. Proceedings of the 19th IEEE Photovoltaic Specialists Conference. IEEE;1987:1484.

[31] Sugiura H., Amano C., Yamamoto A. Yamaguchi M. Double heterostructure GaAs tunnel junction for AlGaAs/GaAs tandem solar cells. Jpn. J. Appl. Phys. 1988;27:269. DOI: 10.1143/JJAP.27.269.

[32] Takamoto T., Yamaguchi M., Ikeda E., Agui T., Kurita H. Al-Jassim M. Mechanism of Zn and Si diffusion from highly doped tunnel junction for InGaP/ GaAs tandem solar cells. J. Appl. Phys. 1999;85:1481. DOI: 10.1063/1.369278

[33] Amano C., Sugiura H., Yamaguchi M. Hane K. Fabrication and numerical analysis of AlGaAs/GaAs

tandem solar cells with tunnel interconnections. IEEE Trans. Electron Devices. 1989;ED-36:1026. DOI: 10.1109/16.24344.

[34] Takamoto T., Ikeda E., Kurita H. Ohmori M. Over 30% efficient InGaP/GaAs tandem solar cells. Appl. Phys. Lett. 1997;70:381. DOI: 10.1063/1.118419.

[35] Sasaki K, Agui T, Nakaido K, Takahashi N, Onitsuka R, Takamoto T. Proceedings of the 9th International Conference on Concentrating Photovoltaics Systems. AIP Conference Proceedings. 2013;1556:22. DOI: 10.1063/1.4822190.

[36] Chu PT, Law DL, Woo RL. Singer S, Bhusan D, Hong WD, Zakaria A, Boisver JC, Mesropian S, King PP, Karam NH. 35.8% space and 38.8% terrestrial 5J direct bonded cells. Proceedings of the 40th IEEE Photovoltaic Specialists Conference. Denver, June 2013. DOI: 10.1109/PVSC.2014.6924957.

[37] Dimroth F, Tibbts TND, Niemeyer M. Four-junction wafer-bonded concentrator solar cell. IEEE J. Photovolt. 2016;6:343. DOI: 10.1109/JPHOTOV.2015.2501729.

[38] Yamaguchi M, Ando K, Yamamoto A, Uemura C. Minority-carrier injection annealing of electron irradiation-induced defects in InP solar cells. Appl. Phys. Lett. 1984;44:432. DOI: 10.1063/1.94756.

[39] Yamaguchi M, Ando K. Mechanism for radiation resistance of InP solar cells. J. Appl. Phys. 1988;63:5555. DOI: 10.1063/1.340332.

[40] Yamaguchi M, Taylor SJ, Yang MJ, Matsuda S, Kawasaki O, Hisamatsu T. Analysis of radiation damage to Si solar cells under high–fluence electron irradiation. Jpn. J. Appl. Phys. 1996;35: 3918. DOI: 10.1143/JJAP.35.3918.

[41] Yamaguchi M, Uemura C. Changes in the electrical properties of GaAs due to electron irradiation. J. Appl. Phys. 1985;57:604. DOI: 10.1063/1.334744

[42] Yamaguchi M, Ando C, Uemura C. Carrier concentration effects on radiation damage in InP. J. Appl. Phys. 1984;55:3160. DOI: 10.1063/1.333316.

[43] Yamaguchi M, Okuda T, Taylor SJ, Takamoto T, Ikeda E, Kurita H. Superior radiation-resistant properties of InGaP/GaAs tandem solar cells. Appl. Phys. Lett. 1997;70:1566. DOI: 10.1063/1.118618.

[44] King RR, Fetzer CM, Law DC, Edmondson KM, Yoon H, Kinsey GS, Krut DD, Ermer JH, Hebert P, Cavicchi BT, Karam NH. Advanced III-V multijunction cells for space. Proceedings of the 4th World Conference on Photovoltaic Energy Conversion. IEEE;2006:1757. DOI: 10.1109/WCPEC.2006.279831.

[45] Yamaguchi M, Luque A. High efficiency and high concentration in photovoltaics. IEEE Trans. Electron Devices. 1999;ED-46:2139. DOI: 10.1109/16.792009.

[46] Yamaguchi M, Masuda T, Arak K, Sato D, Lee KH, Kojima N, Takamoto T, Okumura K, Satou A, Yamada K, Nakado T, Zushi Y, Yamazaki M, Yamada H. Role of PV-powered vehicles in low-carbon society and some approaches of high-efficiency solar cell modules for cars. Energy and Power Engineering. 2020;12:375. DOI: 10.4236/epe.2020.126023v.

[47] Yamaguchi M, Masuda T, Arak K, Sato D, Lee KH, Kojima N, Takamoto T, Okumura K, Satou A, Yamada K, Nakado T, Zushi Y, Ohshita Y, Yamazak M. Development of high-efficiency and low-cost solar cells for PV-powered vehicles application. Prog Photovol. (in press).

Chapter 8

GaAs Compounds Heteroepitaxy on Silicon for Opto and Nano Electronic Applications

Mickael Martin, Thierry Baron, Yann Bogumulowicz, Huiwen Deng, Keshuang Li, Mingchu Tang and Huiyun Liu

Abstract

III-V semiconductors present interesting properties and are already used in electronics, lightening and photonic devices. Integration of III-V devices onto a Si CMOS platform is already in production using III-V devices transfer. A promising way consists in using hetero-epitaxy processes to grow the III-V materials directly on Si and at the right place. To reach this objective, some challenges still needed to be overcome. In this contribution, we will show how to overcome the different challenges associated to the heteroepitaxy and integration of III-As onto a silicon platform. We present solutions to get rid of antiphase domains for GaAs grown on exact Si(100). To reduce the threading dislocations density, efficient ways based on either insertion of InGaAs/GaAs multilayers defect filter layers or selective epitaxy in cavities are implemented. All these solutions allows fabricating electrically pumped laser structures based on InAs quantum dots active region, required for photonic and sensing applications.

Keywords: GaAs, heteroepitaxy, photonics, Si, integration

1. Part 1: GaAs growth on Si(001)

The growth of polar zinc-blende GaAs on a non-polar Si(001) substrate can lead to planar defects named antiphase boundary (APB). The APB planes are made of III-III or/and V-V bonds that can propagate in the layer through the {110}, {111} or higher index planes (**Figure 1**) [2, 3].

The elastic strain field due to APB changes atomic distances and hence electronic states, acts as a carrier diffusion and/or non-radiative recombination centers. APBs nucleate at the edges of the single-layer steps present at nominal (001) silicon surfaces. Until now, the APBs formation was mainly inhibited by using (i) off-axis Si(001) substrates tilted by 4–6° towards [110] direction [4, 5] or (ii) Si(211) substrates [6] where Si double-layer steps could be formed easily. However, these wafers are not compatible with industrial Si CMOS standards which uses nominal Si (001) wafers, i.e. with a miscut angle equal or lower than 0.5°. The best option to prevent the APBs nucleation is to promote double layer step formation on nominal Si(001) substrate as it is the case for off-axis wafers.

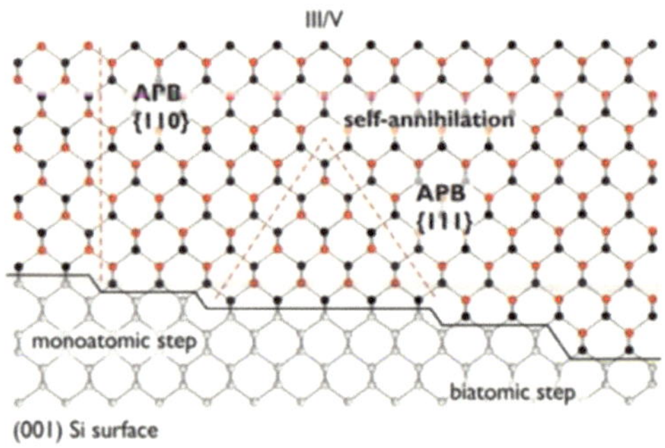

Figure 1.
Ball-and-stick model of III-V-on-Si with {110} and {111}-APBs. The single-layer step edges initiate the formation of the APBs while the surface with double-layer steps allows a single-domain III-V cristal. From reference [1].

1.1 Silicon surface preparation under H_2 ambient

Considering the thermodynamical models, the double-layer steps formation on nominal Si(001) is predicted as highly unfavorable in ultra-high Vacuum (UHV) or in inert gas ambient. The stress relaxation induced by dimerization of the (2×1)-Si (001) reconstruction promotes single step formation until a miscut angle lying in a range between 1 and 3° [7–10]. Thus, the (001) surface of nominal wafers is made of alternating (2×1) and (1×2)-reconstructed terraces (named A-type and B-type terraces respectively) separated by S_A and S_B single steps according to the Chadi's nomenclature (**Figure 2**) [8]. For A-type terraces the Si-dimer rows are parallel to the step edges while they are perpendicular for B-type terraces.

However, computational modeling highlights a possible mechanism to get a nearly single-domain Si(001) surface by selective etching of S_B steps (i.e. by removing the B-type terraces) **under very specific H_2 annealing process conditions** [11]. The energy needed for this process to occur has been calculated by using DFT and the most favorable mechanism has been highlighted. For calculation simplification, a Si(001) 2×2 reconstructed surface has been considered, but the conclusions will be transferable to the real case ie. Si(001) 2×1 reconstruction.

The removal of two neighboring silicon atoms from the considered Si(001) surface (**Figure 2**) creates the so-called single-dimer vacancy (SDV) [12]. Line defects on the surface can appear by aligning SDV together, either in lines, creating dimer-vacancy lines (DVL) or in rows creating dimer-vacancy rows (DVR) [13] as shown respectively in green and pink areas of **Figure 3a**.

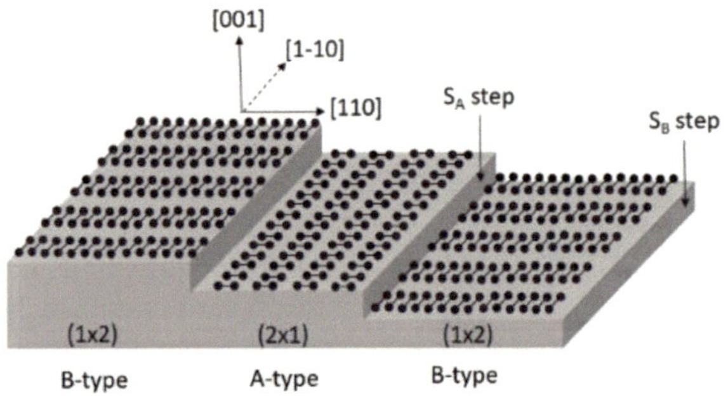

Figure 2.
Dimerization of the Si(001) surface with alternating (2×1) and (1×2)-reconstructed terraces (named A-type and B-type terraces respectively) separated by S_A and S_B single steps.

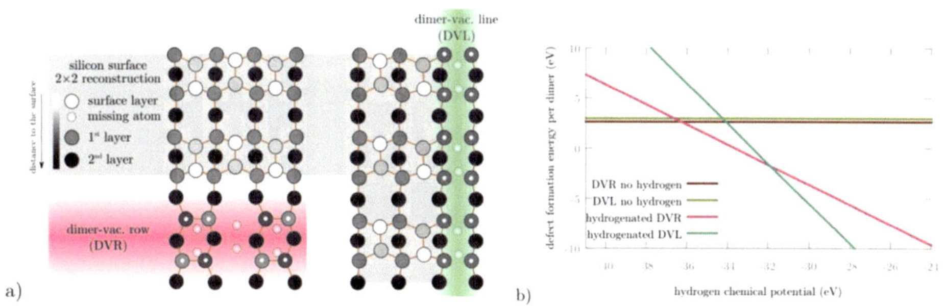

Figure 3.
(a) Schematic view of the DVR and DVL line defects on the 2 × 2 reconstructed silicon surface. Only two silicon bulk layers are represented (black and dark grey) in addition to the surface layer which is reconstructed. The distance from the surface is coded in gray-scale. Silicon atoms marked with a small white disk are hydrogenated in the case of hydrogenated DVR and DVL. (b) The variation of the formation energy of both DVR and DVL defects, bare or hydrogenated with respect to the chemical potential of the hydrogen. From [11].

The formation energy of the bare line defects DVR and DVL (i.e. with dangling bonds, labelled DVR and DVL no hydrogen), are represented on the **Figure 3b**. In order to take into account the hydrogen of the chamber in CVD ambient, the bare line defects are modified by placing a single hydrogen atom on each silicon of the first bulk layer with a dangling bond (**Figure 3a**), changing their geometry and their formation energies. Indeed, the DFT calculations show that for both defects and whatever the surface states, the geometry distortions of the bare defects are considerably reduced when the defect is hydrogenated. In the DVR case, for instance, the dimers in the line adopt a flat position instead of a tilted one. This reduction in elastic stress is key in the formation energy lowering of the line defects. As shown on **Figure 3b**, for hydrogen rich conditions (right handside of the graph **Figure 3b**), the formation energies of both hydrogenated DVR (H-DVR) and DVL (H-DVL) are lower than for the bare defects. Moreover, two regimes can be observed whatever the surface state is. One range for superhigh H chemical potentials where the H-DVL is favored, and a medium range of H chemical potentials where the H-DVR takes prominence. It is worth to note that the gain in energy per dimer can be quite important (several eVs) when comparing the different ranges, showing that the selectivity with respect to H chemical potential is quite strong.

The role of hydrogen is thus twofold. It first induces a large increase of dimer-vacancy concentration due to the lowering of their formation energy. The second effect of hydrogen is to select DVR with respect to DVL when using suitable hydrogen annealing conditions. This latter point is the key to obtain a single domain Si(001) surface. Indeed, the DVRs cross the B-type terraces in a direction perpendicular to the step edges. This can generates a nearly complete etching of the S_B steps if the terraces are not too large. This assumption was tested with a 600 Torr/900°C/10 min H_2 annealing of different on-axis Si(001) wafers. Prior to the H_2 annealing, the native oxide is removed by SICONI™ process [14]. Atomic force microscopy (AFM) images of **Figure 4a**. shows the result obtained from a wafer with a very slight 0.05° misorientation near the [110] direction. The DVRs that run across the B-type terraces can be clearly distinguished. They lead to comb-like shaped B-type terraces. Nevertheless, the terrace width is too large (the miscut angle too small) to obtain a complete removal of B-type terraces. On the contrary, when using a wafer with an higher misorientation (0.15°) in the [110] direction, the B-type terraces can be selectively etched as shown in **Figure 4b**. The AFM line profile confirms the formation of double steps (~2.7 Å in height)). However, there is still a few small islands remaining at the step edges (not clearly visible from the AFM image). This behavior was also observed by other authors working on

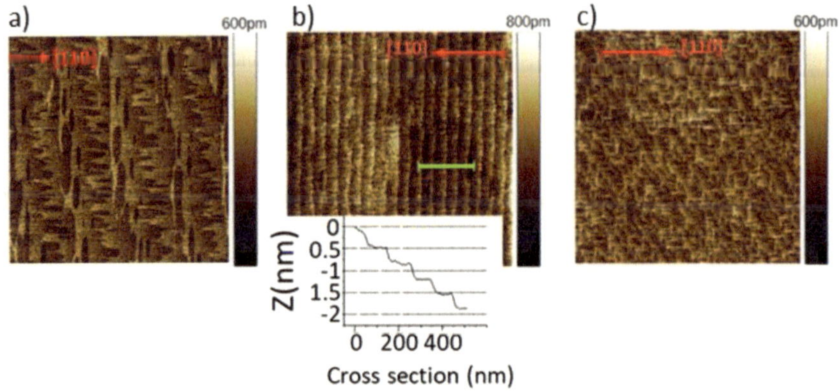

Figure 4.
2 × 2 μm² AFM image of nominal Si(001) surfaces after 600 Torr/900°C/10 min H_2 annealing (a) substrate 0.05° misoriented near [110]. DVRs crosse the B-type (A-type) terraces in a direction perpendicular (parallel) to the single-step edges (b) substrate 0.15° misoriented in [110]. The surface is double-stepped with terraces width of ~100 nm and step height of ~2.7 Å (line profile in inset). (c) substrate 0.12° misoriented near [100]. H-DVRs lead to dendritic single-steps oriented in <110> directions. Images partially reproduced from [11].

GaP-on-Si growth [15, 16]. H_2 annealing of a Si(001) substrate with a 0.12° miscut near the [100] direction was also tested. When the miscut direction slightly differs from the <110> azimuthal directions, each terrace boundary is made of two types of step edge (with both S_A and S_B steps). The selective etching of the S_B-segments leads to a dendritic shape of the terraces (**Figure 4c**). Thus, it is not possible to achieve double-layer steps formation on wafers having a miscut direction different from <110>. It should also be noted that the S_B step etching only occurs for hydrogen conditions near the atmospheric pressure and for a temperature > 850°C [11]. Otherwise, the generation of dimer-vacancies agglomeration is energetically not favorable for such hydrogen chemical potential (**Figure 3b**).

1.2 APBs-free GaAs growth on Si(001)

The effectiveness of the Si surface preparation for APBs removal was proven from GaAs growth on different types of nominal wafers. MOCVD growth of GaAs-on-Si can be achieved using a two-step process [4, 5]: few nanometers of a high-density nucleation layer is first deposited at low temperature (350–450°C) followed by the coalescence of the nuclei during temperature ramp up to 550–700°C. Then at this temperature, a thicker GaAs layer is epitaxially grown to improve the material quality. Classical group-III precursors are TMGa or TEGa while group-V precursors are often TBAs or AsH_3 for the high temperature step. The precursors are injected in the MOCVD chamber using purified H_2 as carrier gas. **Figure 5a** shows the morphology of the GaAs surface grown on Si wafer with miscut angle <0,1° (the type of Si surface presented in **Figure 4a**). Thanks to their V groove shapes, randomly oriented APBs can be observed by AFM with a linear density of several μm^{-1}. This APB density is equivalent to the one obtain for a GaAs growth on a silicon substrate without any surface preparation. It results in a large surface roughness with a root mean square (RMS) value of about 1.5–2 nm. As mentioned before, the APBs originate at the single-step edges between the very large (2 × 1)/(1 × 2)-Si(001) terraces of **Figure 4a**. Thus, the self-annihilation of the APBs is not possible in the GaAs layer (with a typical thickness around 400 nm) due to the large inter-APB distance.

As the Si wafer miscut angle is increased above 0.1° exactly in the [110] direction, the APBs can be easily removed. The AFM image of a 150 nm thick GaAs layer

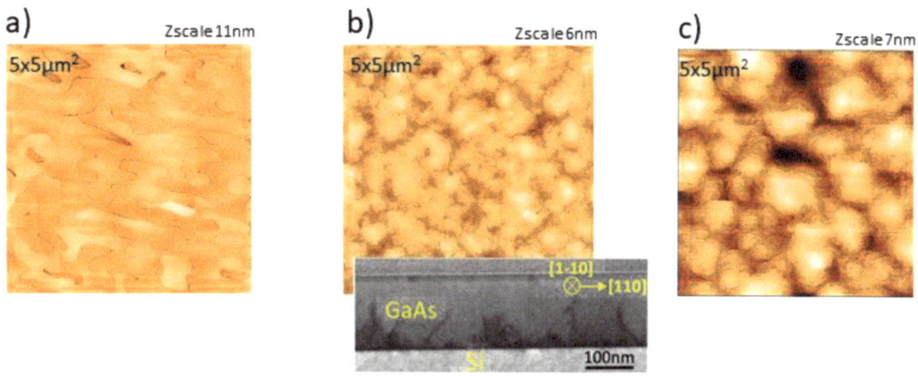

Figure 5.
5×5 µm² AFM images of (a) 400 nm-thick GaAs epitaxially grown on 0.05°-miscut angle Si(001) wafer: High density of randomly oriented APBs. RMS roughness = 1.7 nm. (b) 150 nm-thick epitaxially grown APBs-free GaAs on Si(001) wafer with a 0.15°-miscut angle toward the [110]. The (110)-STEM cross section in inset shows a layer free of APBs beyond about 70 nm of thickness. (c) APBs-free GaAs film grown on Si wafer with a > 0.1°-misorientation toward a random direction. In this case, 300-400 nm of thickness is necessary to get rid of APBs. Images partially reproduced from [11, 17].

is shown in **Figure 5b**. The surface roughness is improved and the RMS value drop to 0.8 nm. This roughness is similar to the one reported for 1 µm thick GaAs grown on 4°-6° offcut Si(001) substrate [18–20], despite the fact that only 150 nm of GaAs were grown. No V-groove feature is observed indicating that a APBs-free surface is formed. The absence of APBs on top of the GaAs layer is confirmed by the STEM cross section image (inset of **Figure 5b**.). The (110)-STEM cross-section shows a dark zone at the bottom of the GaAs layer due to the highly defective Si/GaAs heterointerface. The defective area is a combination of multiple crystalline defects such as dislocations, stacking faults and APBs due to the remaining small monoatomic silicon islands mentioned before. However, for a thickness beyond about 70 nm, no more APB planes propagate toward the surface. Indeed, the APB planes that nucleate at these monoatomic step edges have intersected pairwise during the high temperature growth and thus self-annihilated. The kinking of APBs in the III-V layers followed by their self-annihilation is often explained by kinetic phenomena [21, 22]. In such mechanisms, the adatoms incorporation rate is anisotropic along the two azimuthal <110> directions. In GaAs, several groups have indeed already shown a diffusion constant of Ga atoms 4 times larger along the As dimer lines regarding to the one along the dimer rows [23–25]. It results in a bias between the growth rate of the domains in antiphase responsible for the kinking of the APBs.

Interestingly, a GaAs film grown on Si wafer with a misorientation above 0.1° and toward a random direction (different from the <110>) is also APBs-free beyond a thickness of about 300-400 nm (**Figure 5c**). This can be achieved even though the Si surface is only made of single-layer steps. Actually, this silicon surface, described in the previous section and in **Figure 4c**, is made of very narrow dendritic terraces. Thus, the $(2 \times 1)/(1 \times 2)$-Si domains size are small enough to enable the self-annihilation of the APBs. However, when the misorientation is not in the <110>, a 100 nm-thick GaAs layer is not sufficient to get rid of the APBs and a thicker buffer layer is required.

In an industrial point of view, this is particularly important to relax the constraint on the wafer miscut specifications. Any substrate with a miscut-angle >0.1° can be used whatever the in-plane direction of the wafer slicing. Therefore, contrary to the GaP-on-Si system, the double-layer steps on nominal silicon wafers is not mandatory to achieve a APBs-free GaAs layer.

In the same fashion, the GaAs layer can be epitaxially grown by using a lattice-matched Germanium buffer layer [26]. Beyond the APBs issue, using a relaxed Ge buffer layer is also an interesting strategy to decrease the threading dislocation density in the GaAs layer, as we will see in the next paragraph. Due to the fact that GaAs will be quasi-lattice matched to the Ge strain relaxed buffer and thanks to a reuse of the existing threading dislocations to create new misfit sements if needed, the GaAs/Ge interface should exhibit no such high density of defects as when growing directly GaAs on Si. This will permit a clearer view of the GaAs/Ge interface to precisely observe the defects present when growing polar material on Ge, which is not possible when growing GaAs directly on Si.

5×5 μm^2 AFM images (**Figure 6**) show the surface of 300 nm thick GaAs layers grown on 1 μm-thick Ge/Si(001) substrates with three different offcut angles in the <110> direction: (a) 0.1°, (b) 0.3°, and(c) 0.5°. The APBs density decrease as function of the miscut angle. With a silicon wafer having a miscut angle of 0.5° the 300 nm-thick GaAs layer is completely free of APBs.

Contrary to the direct growth of GaAs on Si, we still observe APBs with a 0.3° offcut Si substrate.

In order to have insight on the defects at the interface in this case, cross-sectional TEM images of a GaAs layer grown on Ge-buffered Si substrate with a 0.5° offcut in the <110> direction are shown in **Figure** 7. The left hand image shows the overall stack, with (from bottom to top) the 0.5° offcut silicon substrate, the 800 nm thick Ge strain relaxed buffer and the 280 nm thick GaAs layer. The interface between GaAs and Ge is highlighted by a thin white line superimposed in the left hand part of the image. No APBs nucleating at this interface are observed, but some dark dots are nevertheless present. The image in the right hand part of **Figure 5** is a magnified view of this interface, showing randomly distributed, different size dark dots which are small (<50 nm), and not at the origin of any extended defects.

Higher resolution images of two of these interface defects have shown that dark spots at the interface are voids, not APBs. Therefore, we observe no APB when using a 0.5° offcut Si substrate for growing GaAs with an intermediate Ge strain relaxed buffer. This hints that bi-atomic steps are achieved at the surface of the Ge strain relaxed buffer using the appropriate hydrogen bake ($T > 750°C$ at 80 Torr H_2), and we observe no APB annihilation such a seen previously when growing GaAs directly on Si.

The progressive improvement of the GaAs layer quality as function of the Si-miscut angle can also be observed from the FWHM of the (004) diffraction line in XRD ω-scan (**Figure 8**).

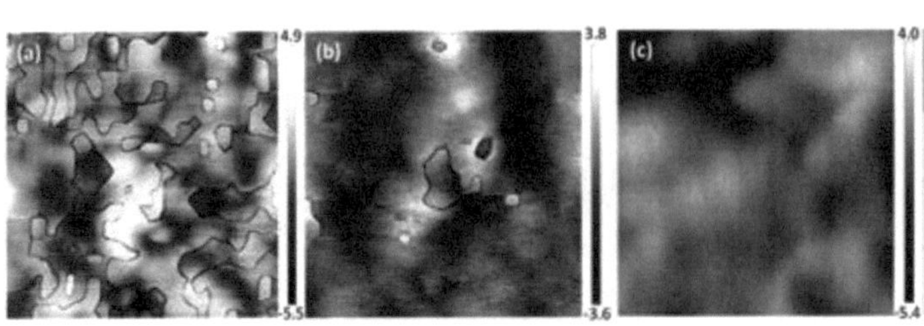

(d)

Sample	Miscut (°)	APB density (μm^{-1})
a	0.1	2.8
b	0.3	0.3
c	0.5	0

Figure 6.
5×5 μm^2 AFM images of the surface of GaAs layers grown on Ge-buffered silicon(001) substrates with three different offcut angles: (a) 0.1°-offcut angle, (b) 0.3°-offcut angle, and (c) 0.5°- offcut angle. All the offcut angles are in the <110> direction. The scale on the right hand side of each image is labeled in nm. The table (d) presents the APB density measured for each sample. AFM image sides are along the <100> directions. From [26].

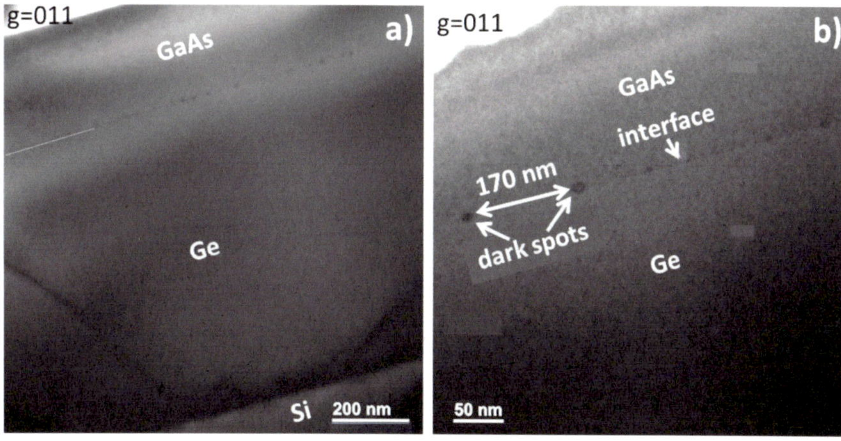

Figure 7.
Cross-sectional TEM images of a GaAs layer on a Ge-buffered Si substrate. The left hand image is an overview of the overall stack, with the 0.5° offcut Si substrate at the bottom. The right hand image is a zoom of the GaAs/ Ge interface.

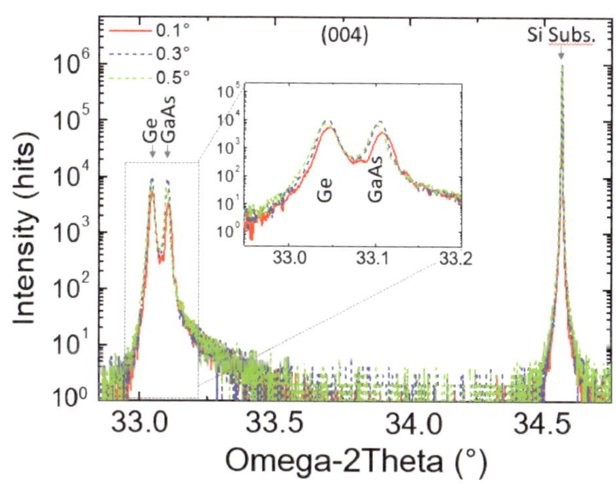

Figure 8.
High resolution, X-ray diffraction profiles around the (004) order (in the triple axis configuration) for a GaAs layer grown on a Ge-buffered silicon substrate with a 0.1° (solid line), 0.3° (dashed line), and 0.5° (dotted line) offcut. From [26].

Detrimental influence of APBs on the optical properties is highlighted from photoluminescence (PL) measurements at 300 K [17]. PL spectra of **Figure 9** compares the near band edge luminescence (1,42 eV) of GaAs-on-Si layers with and without APBs. Both layers are n-doped at 7.10^{17} cm^{-3}. The PL intensity of the APBs-free GaAs film is three times higher than the one of the GaAs layer with APBs. Furthermore, the PL peak of the APBs-free is 40% narrower. These results are directly correlated to the role of APBs acting as non-radiative recombination centers.

In the same way, the influence of APBs on the electrical properties is highlighted from the Hall effect measurements on a 250 nm-thick GaAs active layer n-doped at 7.10^{17} cm^{-3}. This n-doped active layer is grown on intrinsic GaAs-on-Si buffer layers with/without APBs. Hall effect measurements, in the Van der Pauw configuration, are performed by taking 5 points across the whole 300 mm wafer. The mean electron mobilities are reported in **Table 1**. The electron mobility (μ_e) of the GaAs active layer grown on the APBs-free buffer layer is one decade higher than the one

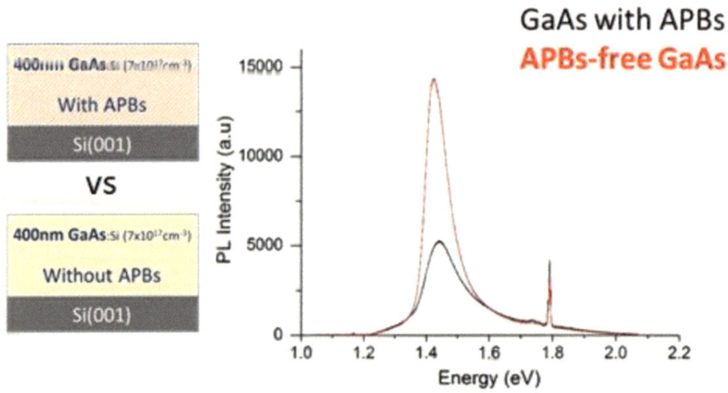

Figure 9.
PL spectra at 300 K for GaAs-on-Si layers with and without APBs. The PL intensity is 3 times higher for the layer without APBs. The FWHM of the peak is 40% lower [17].

250 nm GaAs:Si (7 × 10^{12} cm^{-3}) Active layer without APBs		**VS**	**250 nm GaAs:Si(7 × 10^{12} cm^{-3}) Active layer without APBs**
400 nm GaAs buffer With APBs			**400 nm GaAs buffer Without APBs**
Si(001)		Si(001)	
Sample	Doping level (cm^{-3})	Electron mobility (cm^2/V.s)	Resistivity (W/cm)
GaAs-on-Si with APB	7 × 10^{17}	200	2 × 10^{-2}
GaAs-on-Si without APB	7 × 10^{17}	2000	4 × 10^{-3}
GaAs-on-GaAs	7 × 10^{17}	2500	3 × 10^{-3}

Table 1.
Hall effect measurements at 300 K for GaAs-on-Si layers with/without APBs. The electron mobility is 10 times higher for the layer without APBs [17].

obtained on the buffer with APBs. The μ_e = 2000 cm^{2}V^{-1} s^{-1} value of the APBs-free layer is nearly equivalent to the mobility measured from an homoepitaxy n:GaAs-on-GaAs in the same reactor.

1.3 Summary

APBs formation during the heteroepitaxy of GaAs on nominal Si(100) substrates has hindered for a long time the development of GaAs devices on a Si CMOS platform. With new technologies and processes established by researchers and tools suppliers to control the atmosphere in growth chambers and prepare the Si surface before the epitaxy, one can get rid of APBs easily on GaAs/Si(100).

2. Part 2: reduction of threading dislocations density

The strategies to reduce the threading dislocation density (TDD) in the III-V layer can be classified according to two major tendencies: 1) engineering of thick buffer layer (strained layer superlattices, germanium buffer layer, ...) to annihilate the defects before growth of the III-V active layer. 2) Selective area epitaxy in dielectric cavities (SiO_2,SiN, ...) formed by standard technological steps (deposition, lithography, etching, ...). In this last approach, the threading dislocation (TD) propagation is geometrically limited in one direction by the sidewalls of the patterns. These two majors will be described more deeply in the following paragraphs.

2.1 Insertion of dislocation filters

2.1.1 Introduction to dislocations

The technology of monolithic integration of III-V on Si is of great interest due to combining the superior optical properties of III-V materials and the advantages of Si substrates such as low cost and high scalability [27]. However, as most III-V semiconductor materials have a relative large difference in lattice constant to Si, high density of crystal defects are generated during the epitaxial growth. This leads to the failure of the technique of direct deposition of III/V on Si become commercially viable in 1980s, despite intensive studies have been demonstrated in that era [28]. The lattice mismatch property creates a substantial stress accumulation in the first few pseudomorphic layers of deposited material, as shown in **Figure 10a**. As the stress is accumulated above a critical value of growth thickness, the strain-relaxation process leads the generation of misfit dislocations (MDs). The MDs are associated with missing or dangling bonds along the mismatched interface which are shown in **Figure 10b** [30], thus MDs lie entirely on the growth plane. Since the dislocations cannot be eliminated within a crystal due to energetic reasons, the MDs must either reach the edge of crystal or turn upward through the deposited layers to form TDs. As a result, from the transmission electron microscopy (TEM) shown in **Figure 10c**, TDs seem to extend from the interface of III-V and Si, and go through the epilayer. TDs are likely to be transferred from MDs when the distance to the sample edge is much longer than the distance to the epi-layer surface. Meanwhile, TDs could also transfer to MDs either through dislocation glides, extending the misfit segment beneath it, or in active region during the electron–hole recombination through the phenomenon known as recombination-enhanced dislocations

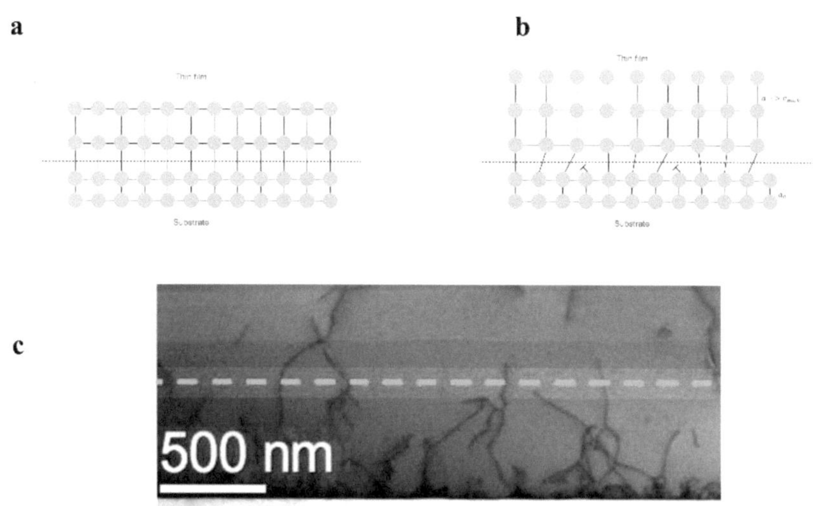

Figure 10.
Schematic change in lattices of thin film on substrates and bright field scanning TEM image of TDs. (a) denoting the initial pseudomorphic layers of deposited materials. (b) denoting MD as spinning "T". (c) a bright-field STEM image showing the TDs. (adapted from Ref. [29]).

motion [31]. Typically, in the growth of GaAs on Si, the TDD is around 10^{10} cm^{-2} at the growth interface [32]. Unfortunately, the viable TDD in active region for practical optoelectronic devices should be below 10^6 cm^{-2} [32], which held back the development of many III-V on Si material systems for some time. Those dislocations have associated trap states serving as nonradiative recombination centers to reduce the photon emission efficiency and/or minority carrier lifetime [32], resulting in a degradation of devices performance. In addition, those states in the band could also increase the leakage current of the devices [3].

Efforts have been made to control the TDD in GaAs grown on Si substrates. As the thickness of deposited layer increases, TDs will glide, move and react with other TDs depending on their Burger vectors, resulting in a repulsion or annihilation as **Figure 11** shows. As TDs keep propagating in the overlayers, they are likely to meet other dislocations to be self-annihilated as shown in **Figure 11**. If there is a strain induced by the lattice mismatched between the underlayer and overlayer, generated TDs are expected to experience lateral forces which drive TDs into edge as **Figure 11** deflection to edge shows.

The deflection process relives the strain induced by lattice mismatch and makes TDs to react with other TDs more likely and/or convert TDs into MDs to decrease the TDD. As demonstrated by Masami and Masafumi in 1990, the dislocation density n in a thick GaAs layer grown on Si can be estimated through the following Equation [34]:

$$n(x) = \frac{1}{\left(1/D_0 + b/a\right) \exp(ax) - b/a}$$

where x is the thickness of the GaAs layer, D_0 is the dislocation density at the interface, a and b are two constants related to the density of etch pit defects (EPD) and coalescence of dislocations respectively. According to their characterization, $D_0 = 10^{12}$ cm^{-2}, $a = 2 \times 10$ cm^{-1}, and $b = 1.8 \times 10^{-5}$ cm. For the dislocation density in a thin GaAs film on Si, it can be estimated through $n(x) = D_0 h^m$, where D_0 is the dislocation density at the interface and m is the empirical value with minus symbol [28, 35, 36]. In order to reduce the TDD to the level of 10^6 cm^{-2}, the thickness of GaAs is estimated to be as thick as 100 μm [34].

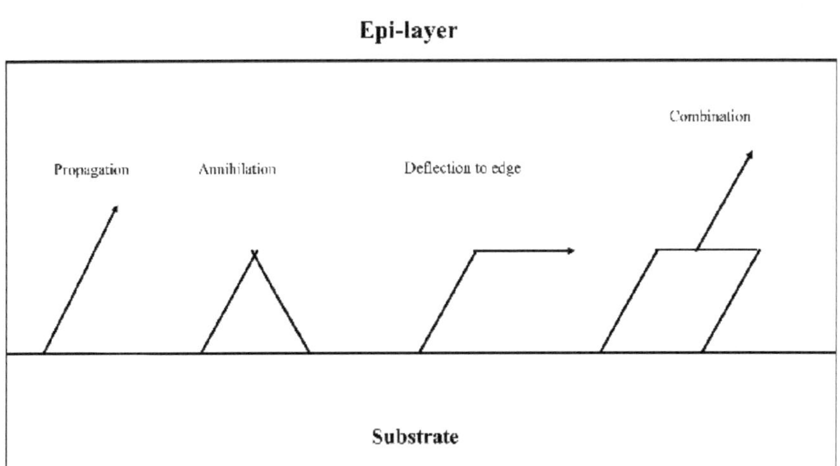

Figure 11.
Mechanisms of dislocation motion in GaAs/Si. (Reproduced from Ref. [33]).

However, due to the difference in thermal expansion coefficients of GaAs and Si, a GaAs layer which is thicker than 7 μm will induce micro cracks on GaAs thin films [37]. In addition, a thick GaAs buffer on Si will bend the wafer [34]. Thus, a more effective method known as dislocation filtering has been put forward to induce designed strain to bend the TDs and to encourage TDs to move, interact and annihilate [28]. The most common dislocation filter layer (DFL) system includes strained layer superlattice (SLSs) and quantum-dot (QD) DFL, while different SLS structure including InGaN/GaN [38], InGaAs/GaAs, InAlAs/GaAs [32] and GaAsP/GaAs [39], have been studied. A typical cross-sectional TEM measurement for InGaAs/GaAs SLSs DFLs is shown in **Figure 12(a)**, indicating how the TDs are eliminated within DFLs. Taking InGaAs/GaAs DFL for example, a layer structure of InGaAs/GaAs is shown in **Figure 12(b)**. The strain direction, induced by the lattice mismatch, inside the DFL is shown in **Figure 12(c)**.

The appropriate choice of composition and thickness for SLS is dependent on the material and the prior dislocation density. Since the purpose of DFL structure is to introduce strain to promote the TD motions, forming dislocations should be avoided within DFL, which means the thickness for each layer should below the critical thickness. For most SLSs, the thickness of each layer should below 20 nm [28, 29, 32]. By using SLS DFL technique, researchers from University College London have successfully reduced the TD density down to the 10^6 cm^{-2} [27].

Although SLSs have been proved to remove more than 90% of TDs [28], the induced strain which bends TDs is still within 2 dimensions. QD is a 0-dimensional nanostructure with much larger strain field compared to SLS. As a result, it is believed that QD can also sever as the DFL, which might be even superior than SLSs [40]. Researchers from University of Michigan have proved that InAs QDs were the most suitable QD. A fabricated laser structure with InAs QD DFL was demonstrated with a threshold current density of 900 Acm^{-2} [40].

2.1.2 Validation of SLS DFL

$In_xGa_{1-x}As$/GaAs SLSs have been recently studied on the GaAs/Si material platform due to its variable strain force. Since the bending efficiency to TDs depends on the strained induced by the lattice-mismatched, the indium composition, the

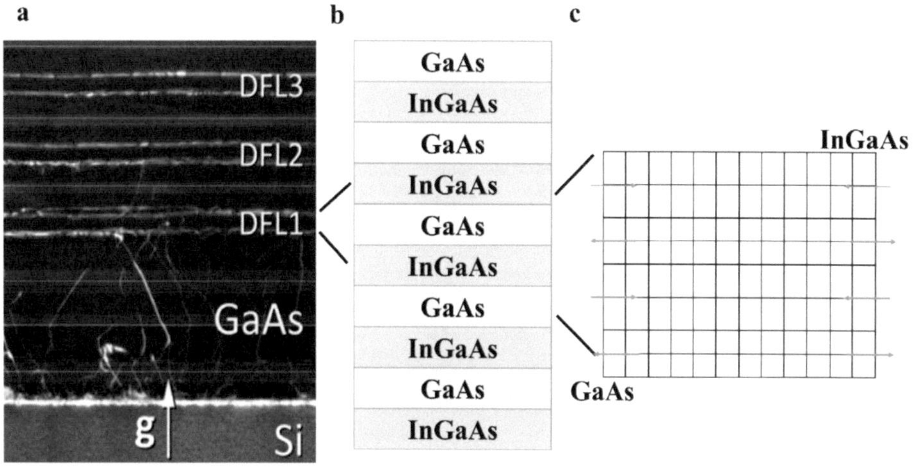

Figure 12.
An InGaAs/GaAs DFL schematic structure in different views. (a) The DFL sample structure in the cross-section TEM. (Reproduced from Ref. [28]). (b) The layer structure of 5 layers of SLSs. (c) The DFL structure in lattice view. The blue arrow denotes the direction of the strain.

thickness of strained layer, and the repetition of SLSs as well as the DFLs are of the greatest interest and need to be considered when optimizing the SLS.

Experiment Techniques

We have investigated the $In_xGa_{1-x}As/GaAs$ SLSs DFLs [32, 41]. As shown in **Figure 13a**, a 1 μm two-step grown GaAs were grown on n-doped Si substrate (001) with 4° offcut towards <011> by using Molecular Beam Epitaxy (MBE) system, while the Si substrate was performed at 900°C for 30 minutes to deoxidize. Then three sets of DFLs were grown, while each DFL structure was composed of five periods of $In_xGa_{1-x}As/GaAs$ SLSs. On the top of DFL, an optimized InAs dot-in-a-well (DWELL) structure was embedded between two 100 nm GaAs layers and 50 nm $Al_{0.4}Ga_{0.6}As$ layers [42, 43]. A final 300 nm GaAs was deposited on the whole structure.

Effectiveness of DFL composed of SLSs

PL measurement was applied through a 635-nm solid-state laser excitation at room temperature. The PL results of InAs QDs with different $In_xGa_{1-x}As/GaAs$ DFL parameters are summarized in **Table 2** and shown in **Figure 14**. In view of the PL intensity and the full width at half maximum (FWHM), it can be known that sample 1 (18% In composition) exhibits a higher PL intensity than sample 2(16% In composition) and sample 3(20% In composition). The PL intensity is highly related to the crystal quality, which corresponds to the TDD. Thus, 18% In composition is proved to contribute to the best crystal quality compared with the other indium compositions. When the thickness of GaAs layer was changed, sample 4 is similar to

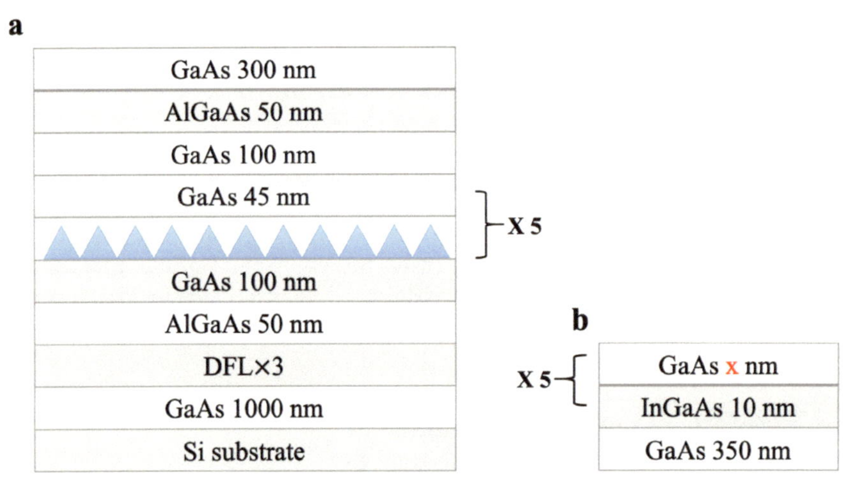

Figure 13.
Schematic diagram of InAs/GaAs DWELL structure monolithically grown on Si substrates with different DFL structures. (a) Schematic diagram of the whole structure. (b) Schematic diagram of the InGaAs/GaAs SLSs as DFL.

Sample	$In_xGa_{1-x}As$	GaAs Thickness (nm)	PL intensity (a.u)	FWHM (nm)
1	$x = 0.18$	10	4	40.3
2	$x = 0.16$	10	2.6	39.8
3	$x = 0.20$	10	2.2	42
4	$x = 0.18$	9	3.9	40.4
5	$x = 0.18$	8	2	46.1

Table 2.
Details parameters for $In_xGa_{1-x}As/GaAs$ SLSs in each sample.

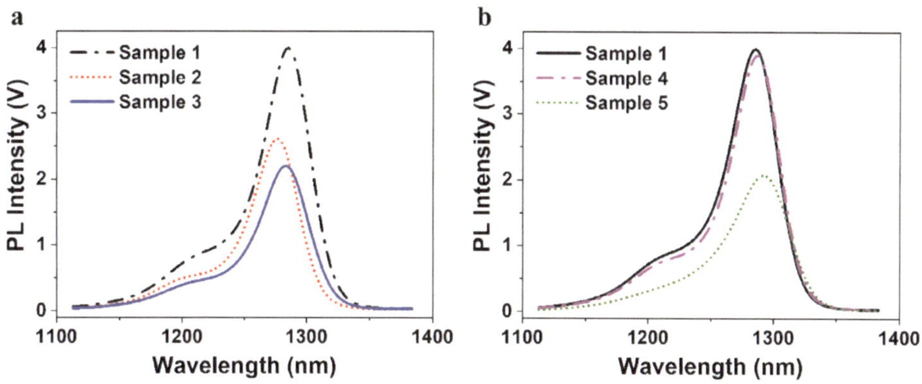

Figure 14.
PL spectra of different samples at room temperature. (a) Sample 1, 2, and 3 with indium composition of 18%, 16%, 20% respectively. (b) Sample 1, 4, and 5 with GaAs spacer layer thickness of 10 nm, 9 nm and 8 nm respectively. (Reproduced from Ref. [32]).

sample 1 in view of the PL intensity, while sample 5 has a significant reduce. In addition, the FWHM of sample 5 is much wider than other 2. This phenomenon is explained through an 8 nm thin GaAs layer cannot release the strain completely so that the accumulated strain degrades the material quality [32].

To further investigate the effectiveness of DFL, Cross-sectional TEM measurements were applied to examine the crystal quality and the effectiveness of DFL. The dark-field TEM image and the bright-field TEM image are shown in **Figure 15**. As shown in **Figure 15a**, high number of TDs appear at the GaAs/Si interface propagating towards the epilayers, as most of them annihilate with others in the first 200 nm. However, there are still a great number of TDs propagating towards the upper layer. After the DFL, only a few TDs puncture the DFL and keep propagating upwards, while most TDs are blocked by the DFL.

In order to further investigate the DFL performance, DFL efficiency (η) is defined as the fraction of TDs it removes, which can be described as [28, 44].

$$\eta = 1 - \frac{n(\text{experiment})}{n(\text{predict})}$$

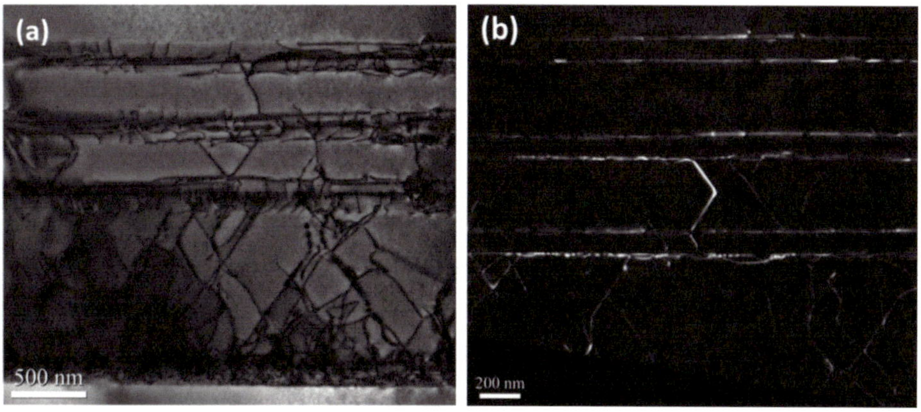

Figure 15.
The cross-sectional TEM image for TDs around DFL structure. (a) Dark-field TEM image (b) Bright-field TEM image (Reproduced from Ref. [32]).

Where n(experiment) denotes the number of dislocations just above the DFL, and n(predict) denotes the dislocations predicted by the equation $n(x) = D_0h^m$, where $m = -0.5$. The efficiencies of different types of DFL are shown in **Figure 16**. From the figure, it can be known that almost half of TDs propagate through the first set of DFL regardless of indium composition of $In_xGa_{1-x}As$ layer. However, the sample with $In_{0.18}Ga_{0.82}As/GaAs$ SLSs shows a superior ability in filtering efficiency compared to others, which achieves over 80%. The demonstrated highest efficiency presents a good balance between TD generation and strain induced to annihilate TDs.

Apart from the InGaAs/GaAs SLSs, InAlAs/GaAs SLSs is also another great option severing as DFL [45–47]. Due to the larger shear modulus of InAlAs, it is expected that the critical misfit for generating new TDs is much larger than that of InGaAs. We compared $In_{0.15}Ga_{0.85}As/GaAs$ DFL and $In_{0.15}Al_{0.85}As/GaAs$ DFL by growing InAs/GaAs QD samples on Si (100) substrates [41]. The $In_{0.15}Ga_{0.85}As/GaAs$ DFL composed three repeats of 5 period of 10-nm $In_{0.15}Al_{0.85}As$ and 10-nm GaAs SLS separated by 400 nm GaAs spacer layer, while the $In_{0.15}Al_{0.85}As/GaAs$ DFL had almost same structure except replacing the $In_{0.15}Ga_{0.85}As$ to $In_{0.15}Al_{0.85}As$. EPD were counted for both samples. After three sets of SLSs, the defects density of the sample with InAlAs/GaAs DFL was around 2×10^6 cm^{-2}, while the other one with InGaAs/GaAs DFL was round 5×10^6 cm^{-2} [41]. In addition, the sample with InAlAs/GaAs DFL had a higher PL peak intensity as well as thermal activation energy compared to the sample with InGaAs/GaAs [41].

2.1.3 Self-assembled QD as DFL

Since it is the Peach-Koehler force in strained layer to bend the TDs to encourage annihilation, self-organized QDs possess an even stronger Peach-Koehler forces, which means QDs are expected to bent TDs more efficiently [48]. Meanwhile, the strain field surrounding QD is 3 dimensions which is superior than 2 dimensions in SLSs.

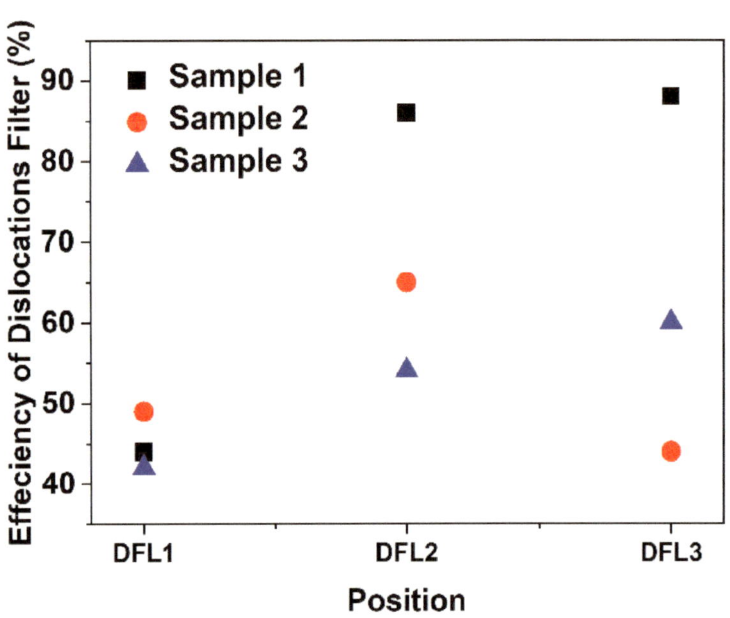

Figure 16.
Summary of the efficiency of dislocation filter for Sample 1, 2 and 3 respectively (Reproduced from Ref. [28]).

Theoretical models.

Several parameters need to be considered when using self-organized QDs as DFLs including QD composition, size, areal density, and the number of dots. The theoretical simulation of the effectiveness of dislocation bending is developed by J. Yang et al. [48], which assuming QD islands are coherently strained with pyramidal in shape. The energy ΔE_{rel} releases when the MD formed to bend the TDs can be calculated through

$$\frac{\Delta E_{rel}}{L} = \frac{2G_{dot}(1+\nu)}{(1-\nu)} f_{eff} b_{eff} h$$

While the dislocation self-energy ΔE_{dis} can be described as

$$\frac{\Delta E_{dis}}{L} = \frac{1}{2\pi} \frac{G_{buff} G_{dot}}{G_{buff} + G_{dot}} b^2 \left(\frac{1 - \nu \cos^2 \beta}{1-\nu} \right) \left[\ln\left(\frac{2r}{b}\right) + 1 \right]$$

The bending will occur when the $\Delta E_{rel} \geq \Delta E_{dis}$. Here, L is the length of the MD, G_{dot} (G_{buff}) is the shear module of dot (buffer layer), ν is the Poisson ration,b is the Burger's vector, b_{eff} is the project of Burger's vector on the buffer layer, f_{eff} is the effective lattice mismatch between the QD and the underlying buffer layer, h is the height of QD, β is the angle between the Burger's vector and the dislocation line, r denotes an outer cutoff radius of the dislocation strain field.

According to the simulations, the bending area ratio, which denotes the bending area divided by the area of QD bases, is shown in **Table 3**. InAs QDs are proved to be the most suitable self-organized QD serving as dislocation filters with the largest bending area and largest critical layer numbers for QD multilayers [48].

Experiment Techniques

Considering the theory and results above, InAs QDs DFL has the highest efficiency compared with other QD DFLs for GaAs monolithically grown on Si. In order to investigate the TD behavior in QD DFL region, a buffer structure shown in **Figure 17** is grown with N-type doped InAs QD dislocation filter on Si (001) substrate with 4° misorientation towards [111]. A thin (<2 μm) GaAs layer is first grown by MOVPE with free of antiphase domain. The TDD at its surface is estimated to be (2–5) × 10^7 cm^{-2}. The dislocation filter consists of 10 layers of InAs QD separated by 50 nm GaAs layers. On the top of QD dislocation filter, a 800 nm GaAs is grown.

Effectiveness of DFL composed of QDs

Cross-sectional TEM measurements were applied to investigate the propagation of dislocations in the QD DFLs. Images were obtained with various g, including [2,$\bar{2}$,0], [1,$\bar{1}$,1], and [0,0,4] as shown in **Figure 18a, b** and **c**, respectively. Two different types of TDs can be observed: pure edge dislocation labeled as C and 60° mixed TDs labeled as A and B. It is obvious from the cross-sectional TEM images

Quantum Dots	Dot Density	QD base area	Bending area ratio of a single QD	Bending area ratio of a single layer
Unit	(cm^{-2})	(nm^2)		
$In_{0.6}Al_{0.4}As$	2×10^{11}	27–56	~0	~0
$In_{0.5}Ga_{0.5}As$	5×10^{10}	80–132	< 1%	~0
InAs	2×10^{10}	120–210	80%	10%

Table 3.
Bending area ratios for different QDs (Reproduced from Ref. [48]).

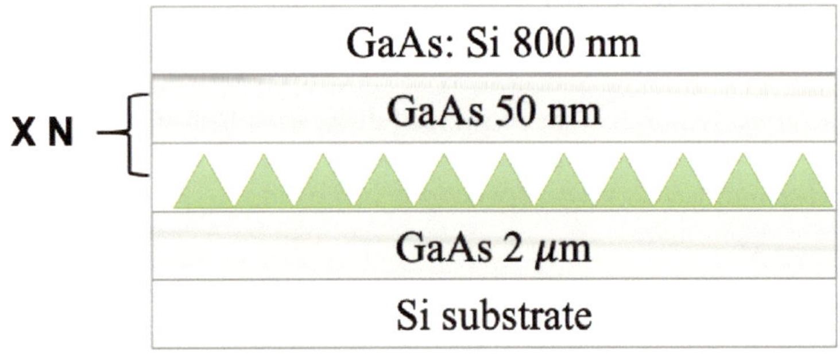

Figure 17.
GaAs grown on Si with 10 InAs QD layer as dislocation filter. (reproduced from Ref. [48]).

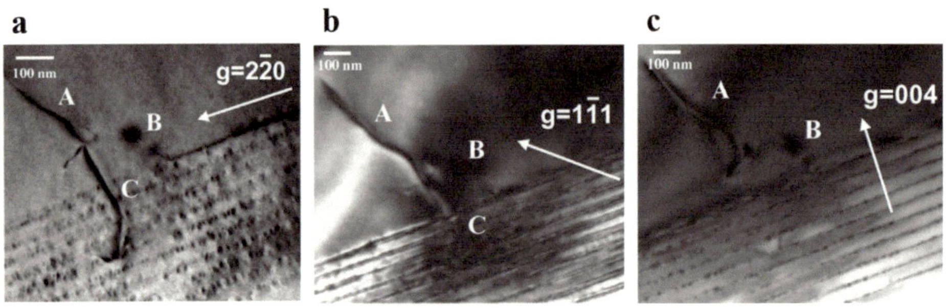

Figure 18.
Cross-section TEM images of dislocation propagation in the ten-layers InAs QD with various diffraction conditions: (a) g = [2,2̄,0], (b) g = [1,1̄,1], (c) g = [0,0,4]. (Reproduced from Ref. [48]).

that the QD DFL can bend 60° mixed TDs effectively. In addition, pure edge TDs, which cannot be blocked by the 2-D SLS [49], can be terminated within the QD DFL. Although the detail of this termination is not fully understood, it is believed that the formation of a dislocation loop or the annihilation with a dislocation with reverse Burger's vector result in the termination [48]. Recently, with the help InAs QD DFL, J. Wang et al. demonstrated a low dislocation density of 2×10^6 cm^{-2} [50], with a high efficiency of 96% calculated.

Conclusion

In the past 30 years, efforts have been made to decrease the TDs induced by the lattice mismatch between GaAs and Si. Many researchers have successfully adopted DFL method to decrease the density of TD to 2×10^6 cm^{-2} [45, 50]. However, a 9-μm thick GaAs buffer is indispensable if no other technique applied to achieve that density. Thus, the DFL technique is much more effective in reducing TDs comparing to grow GaAs buffer, which is summarized in **Table 4**.

With the DFL technique, researchers make it possible to reduce the vast number of TDs to a level which is commercially viable in a thin film around 2.5 μm. This

Technique	Thickness
Thick GaAs buffer layer	9 μm
InAlAs/GaAs SLSs as DFL	2.35 μm
InAs QD as DFL	2.205 μm

Table 4.
Summary of the requisite thickness with different methods to reduce the density of TD to 2×10^6 cm^{-2}.

technique has promoted the implement of III-V materials directly grown on Si such as growth of III-V lasers on Si substrates [27, 50].

2.2 GaAs growth on Germanium strain relaxed buffer

As germanium material has lattice parameter and thermal expansion coefficient close to those of the GaAs, a common strategy is to benefit from all the Ge heteroepitaxy on silicon developments to reduce the structural defects in the GaAs layer [31, 41, 51–53]. This way, we avoid additional threading dislocation nucleation. Currently, the TDD in a 1.5 μm thick Ge-buffer on Si(100) is in the 10^7 cm^{-2} range by using [54, 55] a thermal cycle annealing (TCA). The **Figure 19a**, extract from the works of Bogumilowicz et al. [56], shows the TDD evolution in function of the Ge buffer and GaAs total thickness, with a GaAs layer fixed at 270 nm thick. The GaAs layer is smooth (<1 nm RMS) and free of APBs thanks to the process described in the previous section. The TDD was estimated by using three methods: (i) from the XRD rocking curve width, the value is extracted with the Ayer's model [57] (ii) by counting the dark spots on the cathodoluminescence (CL) image of the GaAs surface, (iii) by counting the pits on the AFM image of the GaAs surface. Whatever the method, the authors show that the TDD tends to reach a plateau at a value around 3×10^7 cm^{-2}. Nevertheless, the downside of the Ge virtual substrate method is the wafer bowing due to the difference between the thermal expansion coefficients (around 120% for Ge and Si). The **Figure 19b** is a plot of the 300 mm wafer bow versus the film thickness. For the thickest Ge buffer layer (1.38 μm) the bow is measured at −240 μm. Such a value is still a hurdle for the wafer handling and processing with the 200/300 mm foundry tools.

2.3 TDD reduction by selective area growth

Selective growth method is often used in heteroepitaxy of semiconductors where cavities are used to block geometrically the propagation of structural defects that generate at the interface of lattice mismatch semiconductors. Different techniques could be implemented such as Epitaxial Lateral Overgrowth (ELOG) and Aspect Ratio Trapping (ART). We will describe more in details the last one.

ART allows to block inside the cavities some of the threading dislocations and planar defects propagating perpendicularly to the trench direction. Still, a few structural defects propagate through the film. **Figure 20a** summarizes the principle

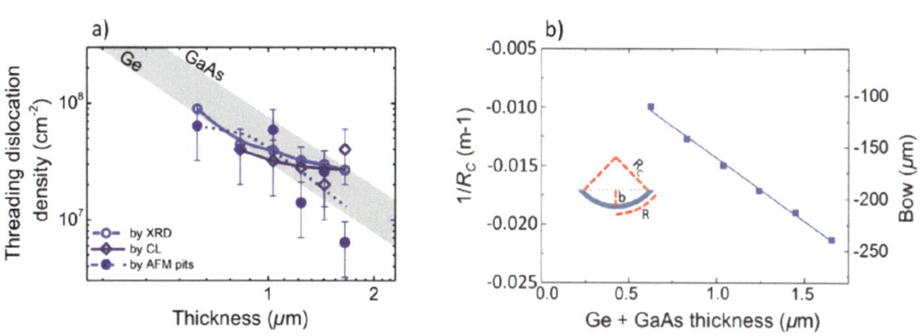

Figure 19.
(a) Plot of the TDD in the GaAs overlayers as a function of the total Ge + GaAs thickness. The light gray area corresponds to the expected TDD values in Ge or GaAs single layers as a function of thickness. Estimated error bars are shown for the TDD extracted from AFM and CL. The TDD error bar for the XRD data is $\pm 10^7 cm^{-2}$. *(b) Plot of the substrate bow versus the total Ge + GaAs thickness.*

of the method. In fact, the TDs propagating through the dense {111}-planes can be blocked by the geometry of the patterns. In this case the aspect ratio (height on width h/l) of the pattern must be such that $h/l \geq \tan\theta_{(111)} = \tan(54.7°) = 1.4$. Therefore, for example, with a 150-thick dielectric, the cavity must be no wider than ~100 nm. Nevertheless, as we can see on the figure, the planar defects which lie parallel to the trenches can be more difficult (or impossible) to trap.

In the example of ART from the works of Lau et al. [58], they use about 100 nm-width oxide trenches on Si(001) etched by a wet solution of KOH to form a "V-groove" (**Figure 20b** and **c**). Indeed, with this type of wet etching the {111}-Si planes are revealed at the bottom of the trenches. This approach has the advantage to free the III-V layer from the APBs which doesn't form on Si(111) surface. The GaAs is then epitaxialy grown to obtain a nanoribbons array in the trenches. The growth process is achieved in a classical way with two steps: one low temperature nucleation (365°C) step followed by a fast growth at high temperature (570°C). The STEM images highlight some microtwins, at the Si/GaAs interface, which are trapped by the V-groove structure (**Figure 20c**). However, outside that thin area the GaAs layer is single-domain with a good crystalline quality. The XRD rocking curve from the (004) peak was measured in both configuration parallel and perpendicular to the line. For a 200 nm-thick GaAs the FWHM of each peaks are measured at 400 arcsec and 550 arcsec for the perpendicular and parallel configuration respectively. That difference is attributed to the defects not trapped by the trenches in the parallel direction. Some of these defects can be seen on the STEM cross-section, parallel to the trench, of **Figure 20d**. The Moiré fringes are formed by the interferences between the Si et GaAs crystal at the V-groove level. Besides the defects trapping by the cavity, the low defectivity is ascribed to the stress relaxation by the partial dislocations associated to the stacking faults and microtwins at the interface. This phenomenon has already been reported with the InP growth in other works [60, 61].

More recently, Kunert et al. [62] used a SAE approach to achieve some GaAs and GaSb nano-ridges (NRs) which come out from the cavities with tunable shapes and facets as function of the process conditions (**Figure 21a**). These type of NRs can serve as laser diodes structure as well as planar photodetectors. The nano-ridges growth is achieved in trenches where the silicon has been etched in wet solution to form a v-groove of Si-{111} planes. The defect density on the top surface of NRs was assessed in the direct space from electron channeling contrast imaging (ECCI). This latter method can be implemented more easily and with a better statistic than TEM. The graph of **Figure 21b** summarizes the TDD and planar defect density (PDD) in

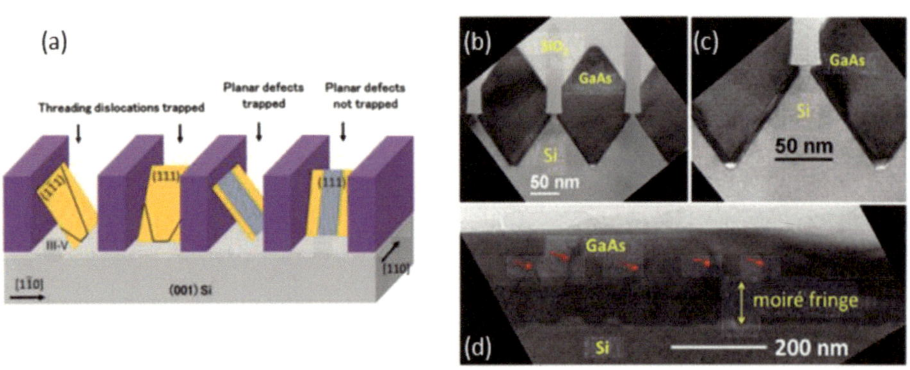

Figure 20.
(a) Principle of ART. (b) GaAs growth in V-groove shaped Si/SiO$_2$ trenches. (c) SFs and microtwins at Si/GaAs interface. (d) STEM cross-section along the trench. The TDs are indicated by the red arrows. The Moiré fringes are formed by the interferences between the Si et GaAs crystal at the V-groove level. From [58, 59].

DOI: http://dx.doi.org/10.5772/intechopen.94609

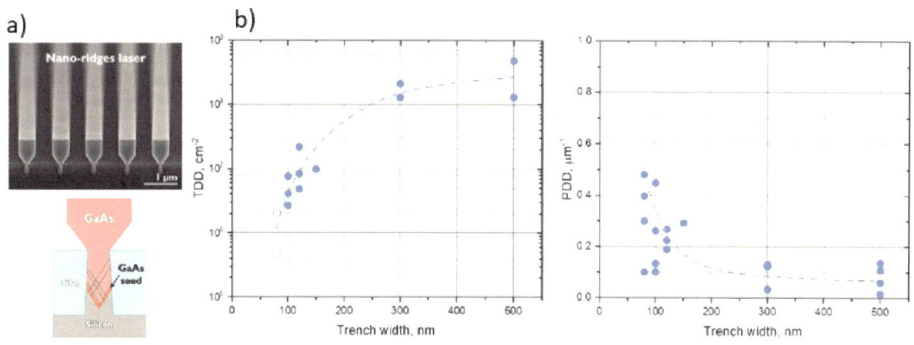

Figure 21.
(a) GaAs nano-ridges in SAE. (b) TDD and PDD as function of the trench-width. From [62].

surface of the GaAs as function of the trenches width. A remarkable achievement is the low TDD which decrease below 4.5×10^{5} cm^{-2} for the very narrow trenches of 80 nm (AR 3.75). Therefore such TDD is therefore very closed to the one of a bulk GaAs substrate (~10^{5} cm^{-2}). This result was observed on both GaAs and GaSb NRs. The planar defect density (essentially stacking faults) shows, however, an inverted relationship with the trench width. The PDD is indeed significantly higher in the narrow cavities. It rises from a value below 0.2 μm^{-1} in the >300 nm-wide trenches to 0.5 μm^{-1} for trenches below 100 nm-wide. The authors assume that the Shockley partial dislocation at the SF-planes ends may help to release the stress in the narrow trenches.

Furthermore, Baron et al. [63] highlighted the efficiency of the ART method for the optical emission of AlAs/InGaAs/AlAs QWs. The QWs are grown on top of a 150 nm-thick GaAs buffer layer in SiO_2 trenches with differents aspect ratio ranging from 0.2 to 1.3 (**Figure 22a**). In this work a 1.3 AR is necessary to free the GaAs buffer layer from APBs and to obtain a PL at 300 K. This way the **Figure 22b** shows the normalized µPL spectra for InGaAs QWs with different Indium content. The PL peak position measurement combined with the InGaAs layer thickness measurement by STEM (**Figure 22c**) allows for the calculation of an Indium content of 7%, 16%, 35%, 42% in the 4 samples. These values are very close to the targeted concentrations. Besides, to observe the influence of defects at the local scale, CL measurements at 15 K were performed on top of the nanoribbon arrays (**Figure 22d**). Since the layer is free of APBs, the dark zone, corresponding to non-luminescent areas, are attributed to the dislocations that are not trapped by the structure and propagating through the QWs.

This explanation of the TDs acting as luminescence quenchers was pushed further in another work [64] combining FIB-STEM, CL and strain measurement of the III-V nanoribbons by precession electron diffraction (PED) [65, 66]. The **Figure 23a.** shows a STEM cross-section of the nanoribbons with their AlAs/InGaAs QWs. The structural defects crossing the QWs are labeled from d_1 to d_5. Prior to the STEM lamella preparation CL intensity imaging (**Figure 23b**) was performed at the same location of the nanoribbon (the area is located thanks to platinum marker deposited on top of the NRs array). The authors highlighted that the luminescence is not homogenous along the NRs and the dark and bright area are bounded by two TDs indexed as d_3 and d_4 on the image. In addition, the CL peak position of the brighter area shift of about 10 nm toward the higher wavelength (**Figure 23c**). Both the intensity and the peak shifting can be spatially correlated to the 0.5% $\varepsilon_{[110]}$ strain variation starting from the d_3 dislocation and measured thanks to the PED method (**Figure 23d**).

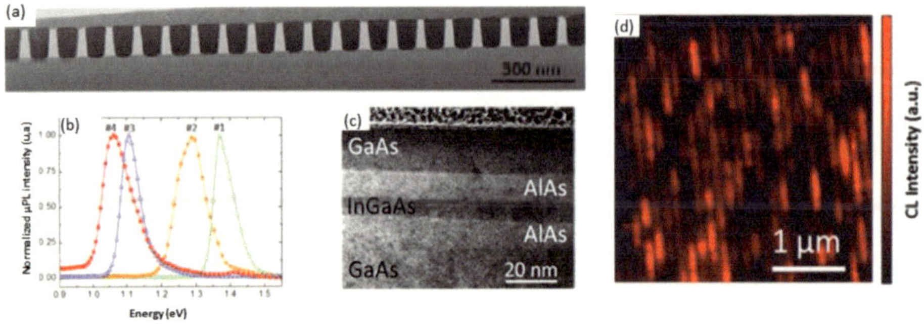

Figure 22.
(a) low magnification cross-sectional STEM image of a GaAs layer grown in 140 nm wide SiO2 trenches on (001)-oriented Si substrate showing a good uniformity of the selective growth. The trenches are oriented along the [1–10] direction and are 180 nm deep. (b) Normalized room temperature lPL spectra of different InGaAs. QWs having different composition of Indium of (#1) 10%, (#2) 20%, (#3) 30%, and (#4) 40%. (c) Cross-sectional TEM image of the top layers showing the stack of GaAs/AlAs/InGaAs/AlAs/GaAs layers with no crystalline defects. (d) 5 K panchromatic CL mapping of the nanoribbons array. From [64].

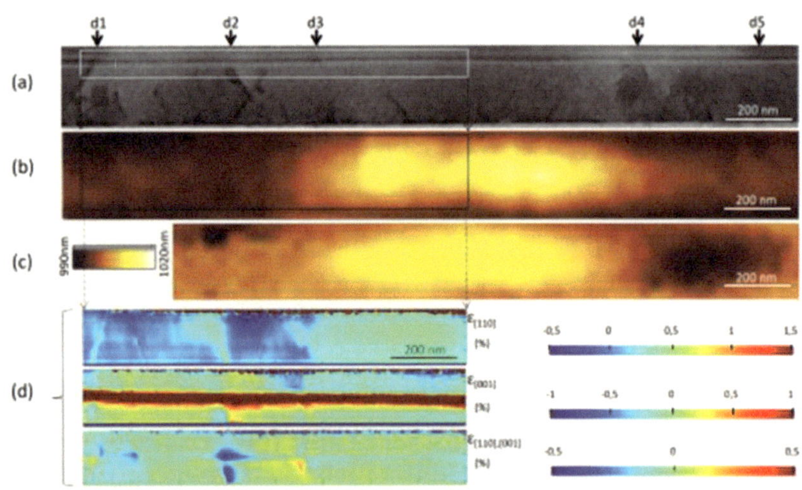

Figure 23.
Spatial correlation between mappings: (a) cross section STEM, (b) top-view CL intensity, (c) CL peak positions, and (d) cross section ε [110], ε[001], and ε [110, 001] strain distortions realized on a single III-V QWF. The high luminescent area is bounded by dislocations d3 and d4 and associated with a peak position shift toward higher wavelength. ε [110] shows a 0.5% distortion along this III-V QWF, and no significant distortion for ε[001] and ε [110, 001]. From [64].

The SAE approach entails a large number of variants, including epitaxial lateral overgrowth ELO [67] and confined epitaxial lateral overgrowth (CELO) [68]. These alternatives often use a "3D" confinement of defects. However, if they are in certain cases, very efficient, they generally require a complex and cost consuming patterning of the substrates. For an overview of the latter methods one can refer to the references [59, 69].

2.4 Summary

In the past 30 years, efforts have been made to decrease the TDs induced by the lattice mismatch between GaAs and Si. Introduction of DFL method as well as the use of Aspect Ratio trapping method allow to decrease the threading dislocation density in the 10^5–10^6 cm^{-2} range, value required to obtain efficient devices.

3. Part 3: realization of InAs QDs/GaAs laser emitter on APB-free GaAs/Si platform

3.1 The Development of QD laser on Si

The advantages of high data rate, broad bandwidth, mature fabrication processes and low power consumption make Si photonics become a desirable approach, meeting the future demands of optical interconnections. To date, significant achievements have been made in Si photonics and most of key components have been well demonstrated, including low-loss waveguides, high-speed modulators and high-performance photodetectors [70–74]. However, the realization of high-performance Si-based on-chip light sources still remains challenging for the full integration of optoelectronics integrated circuits [75]. Among various of approaches, monolithically integrating high-performance III-V QD lasers on Si substrate has been considered as a promising method to develop an on-chip optical source for Si photonics [76–78]. The advanced properties of low threshold, high defects tolerance and high temperature stability contribute largely to the development of QD lasers [27, 79–81].

Before illustrating the recent progress of QD lasers grown on Si (001) substrates, it is worth to discuss briefly about some key milestones in the development of monolithic integration of III-V lasers on Si. Although some optimized heteroepitaxy techniques have reduced the TDD of III-V on Si from originally $\sim 10^9$ cm^{-2} to $\sim 10^6$ cm^{-2}, QW lasers directly grown on Si still suffered on their high threshold and limited lifetime due to the enhanced TD generation [82–85]. An early result presented a QW laser with InP buffer as thick as 15 μm with decent performance and lifetime [86]. However, due to the difference of thermal expansion coefficient between III-V epi-layer and Si substrate, the thick buffer is also vulnerable to the formation of micro-cracks, which destroys the yield of Si-based devices [87]. The research of III-V QD lasers on Si (001) comes out since early 2000s. After the early attempt by using droplet epitaxy to grow QD lasers, the successful address of Stranski-Krastanov growth mode on the growth of QDs presents significant advantages on emitting light with the presence of high TDD caused by mismatch in lattice constants and thermal expansion coefficients [88, 89]. By taking the benefits of ultra-high vacuum and precise control, MBE system has been widely considered as a suitable technique for the growth of high-performance QDs.

Recently, numerous achievements that pursuing high performance III-V QD lasers on Si have been demonstrated. The offcut Si substrate was addressed initially to prevent the formation of APB. In 2001, the first QD laser on Si emitting at 855 nm at room temperature under continuous-wave operation was presented by growing InGaAs QDs on Si substrate with MOCVD [90]. More importantly, the aging test illustrated the advantage of reliability for QD lasers on Si compared with QW counterparts. By further optimizing the active region and III-V buffer, such as utilizing DFLs and P-type modulation doped QD region, the performance of QD lasers on Si was highly improved, realizing a characteristic temperature (T_0) of 244 K between operation temperature of 25–95°C and a reduced threshold current density of 900A/cm^2 at that time [48, 91]. These results suggest the possibility of QD lasers directly grown on Si substrate as an efficient and reliable light source for Si photonics.

The aforementioned works of QD lasers were all operated under emission of 1.1 μm. However, the recent ever-growing demands on telecommunication and data-communication system, led to significant achievements on 1.3 μm InAs/GaAs QD lasers on Si substrate. The first room temperature 1.3 μm emission of QDs on Si grown by MOCVD was achieved by Li et al. at 2008, with the help of Sb [92].

However, due to the high TDD in the GaAs buffer, its PL intensity was eight times weaker than QDs grown on the native GaAs substrate, even with a high QD density obtained of 7×10^{10} cm^{-2}. This also suggested the importance of developing improved GaAs buffer on Si substrate associated with well-performed DFLs. The first electrically pumped 1.3 μm InAs/GaAs QD laser directly grown on Si substrate by MBE was successfully demonstrated by Wang et al. in 2011 [93]. The laser structure was grown on an offcut Si substrate with 4° miscut angle to [110] orientation. An improvement initialized from the growth of AlAs nucleation layer instead of GaAs nucleation layer, realizing a reduction of defects observed at the interface of AlAs/Si [94]. The threshold current density was reduced to 725A/cm^2 at room temperature under pulsed operation, with a single facet output power of ~26 mW achieved at room temperature. The highest operation temperature was 42°C with a T_0 of 44 K.

Extensive studies were devoted following the first demonstration of electrically pumped 1.3 μm QD laser on Si. In 2012, by utilizing Ge-on-Si virtual substrate, the first room-temperature continuous-wave electrically pumped InAs/GaAs QD laser monolithically grown on Si substrate with a Ge buffer layer was demonstrated by MBE [95]. A low threshold current density of 162 A/cm^2 was achieved at continuous-wave mode with a room temperature lasing emission of 1.28 μm. The operation temperature was as high as 84°C under pulsed mode. Although these results were outstanding, a 1.3 μm InAs/GaAs QD laser directly grown on Si substrate was still far from practice, until the successful demonstration by us in 2016. By applying unique epitaxial method and improved fabrication process, the first high-performance and long-lifetime 1.3 μm QD laser directly grown on Si was achieved [29]. As shown in **Figure 24a, 1** μm GaAs buffer was grown by three steps on a deoxidized Si substrate to improve the material quality, followed by four sets of DFLs consisted of five sets of InGaAs/GaAs SLSs and high temperature annealed

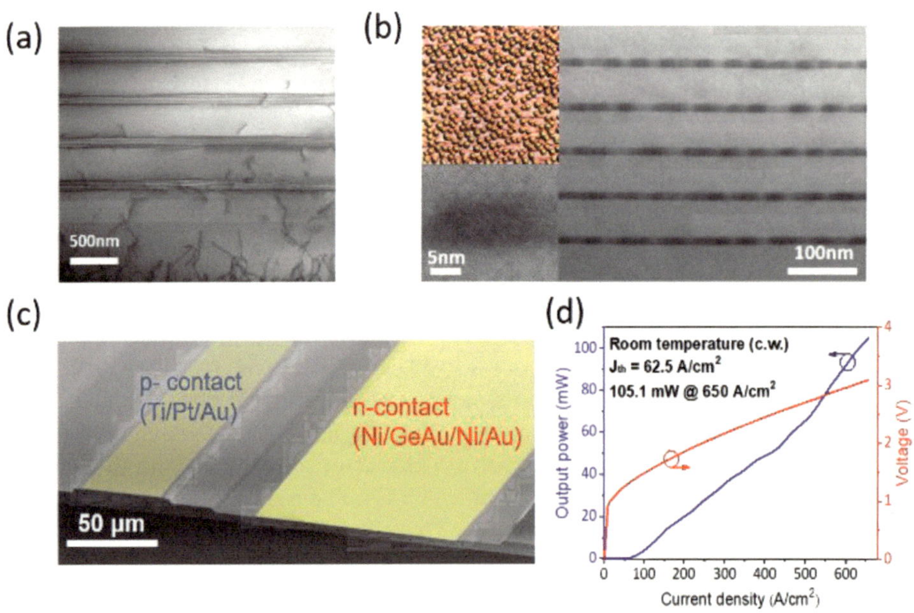

Figure 24.
(a) TEM image of GaAs buffer on Si including dislocation filter layers. (b) TEM image of active region, upper inset: 1 × 1 μm² AFM image of uncapped QDs, and bottom inset: TEM image of a single QD. (c) SEM image of fabricated broad-area laser. (d) Light-current–voltage curve of lasing characteristics under continuous-wave condition at room temperature. Reproduce from [29].

300 nm GaAs spacer layer. The TDD after DFLs was successfully reduced to the level of 10^5 cm^{-2}. High-performance laser structure with five stacks of InAs/GaAs DWELL active region was developed upon this platform. A TEM image of active region was shown in **Figure 24b** where QDs were coherently grown, without any visible defects. The two inset images presented a 1×1 μm^2 AFM image which show a good uniformity with 3×10^{10} cm^{-2} dot density and the typical shape of a single QD. Broad-area lasers were fabricated as shown schematically by a scanning electron microscope (SEM) image in **Figure 24c**. The light-current–voltage curve of the device was shown in **Figure 24d**. An ultra-low threshold current density of 62.5 A/cm^2 under continuous wave at room temperature was obtained, which was the lowest threshold current density value achieved for any kind of lasers on Si substrate at that time. The single facet output power measured under injection current density of 650 A/cm^2 was exceeded 105 mW. The highest operation can be achieved up to 75°C under continuous-wave mode and 120°C under pulsed mode. Moreover, negligible degradation was observed after 3100 h aging test, realizing an extraordinaire mean time to failure lifetime more than 100,158 h. After that, Si-based monolithically integrated narrow-ridge Fabry-Perot and distributed feedback QDs laser are fabricated based on these outstanding outcome [96, 97].

3.2 InAs/GaAs QD laser on on-axis Si (001) substrate

These previous discussions on QD lasers were all fabricated on offcut Si substrate, which are not fully compatible to the CMOS technique. The commercialized on-axis Si (001) platform demands an miscut angle less than 0.5°. As discussed in the first section of this chapter, the heteroepitaxy technique on on-axis Si (001) was satisfied by forming APB-free GaAs buffer. Beyond the successful demonstration of QD lasers on offcut Si platform, QD lasers grown on CMOS-compatible Si (001) substrate were successfully developed in recent years [98–105], owing to the demonstration of the APB-free GaAs and GaP templates on Si.

The first electrically pumped continuous-wave InAs/GaAs QD laser monolithically grown on-axis GaAs/Si (001) substrate was demonstrated in 2017 [98]. Following by a 400 nm APB-free on-axis GaAs/Si (001) platform grown by MOCVD, MBE system was employed to grow QD laser structure with four repeats of DFLs, which consist of five sets of InGaAs/GaAs SLSs. The five stacks of InAs/GaAs QD layers sandwiched by AlGaAs cladding layers were grown subsequently. A 1×1 μm^2 AFM image of uncapped InAs QD on Si (001) substrate was shown in **Figure 25a**, realizing a good uniformity and a dot density of $\sim 3.5 \times 10^{10}$ cm^{-2}. The sample was fabricated to broad-area laser devices with as-cleaved facets for laser characteristic measurements. A comparison of room-temperature continuous-wave light-current–voltage characteristics between QD laser on on-axis GaAs/Si (001) platform and native GaAs substrate was shown in **Figure 25b**. The GaAs-based QD laser presented a threshold current density of 210 A/cm^2, while that of on-axis Si-based QD laser was 425 A/cm^2. The calculated slope-efficiency and differential quantum efficiency of GaAs-based QD laser were $\sim$ 0.12 W/A and 12.7%, respectively. The QD laser on on-axis Si (001) also show decent results on corresponding characteristics, which the calculated slope-efficiency was 0.068 W/A and differential quantum efficiency was 7.2%. **Figure 25c** presents a temperature dependent light-current curve of Si-based QD laser operated under continuous-wave mode. The maximum operating temperature achieved was 36°C. As shown in **Figure 25d**, the pulsed results of light-current characteristic at various heatsink temperature presented a highest operation temperature of 102°C, which was the first demonstration of QD laser directly grown on on-axis Si (001) substrate that observed lasing over 100°C. The inset image of **Figure 25d** shows the characteristic

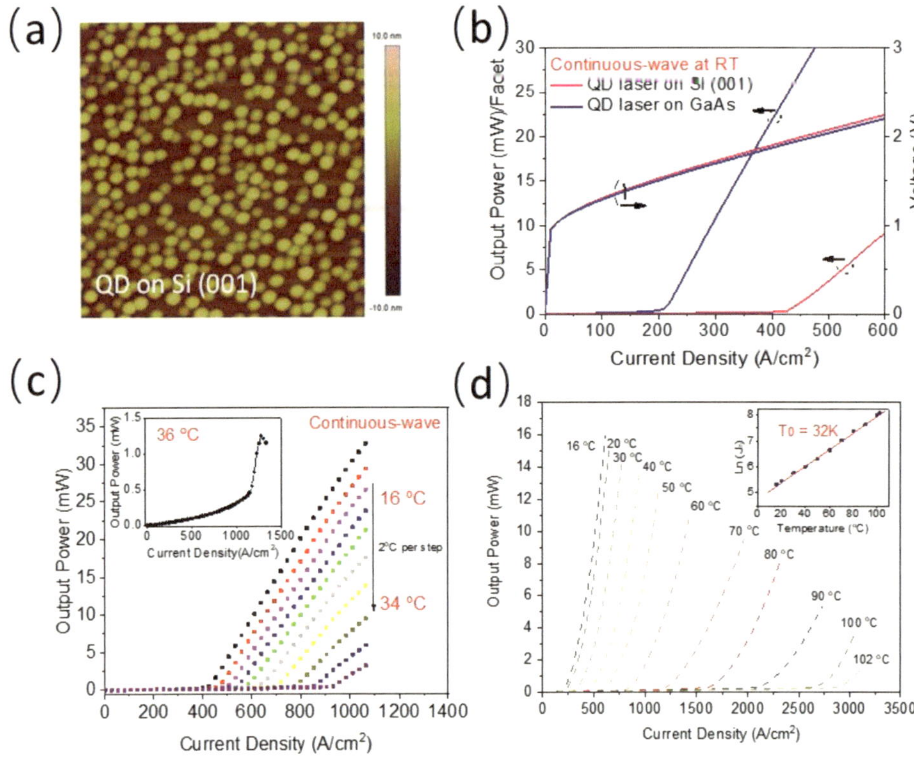

Figure 25.
(a) 1 × 1 μm² AFM image of uncapped InAs QDs grown on on-axis Si (001) substrate. (b) Light-current–voltage characteristic comparison of an InAs/GaAs QD laser grown on on-axis Si (001) and native GaAs substrate at room temperature under continuous-wave operation. (c) Single facet light-current curve for InAs/GaAs QD laser on on-axis Si (001) as a function of temperature under continuous-wave operation, inset: light-current curve at a heat sink temperature of 36°C. (d) Single facet light-current curve for InAs/GaAs QD laser grown on on-axis Si (001) substrate at different heat sink temperatures under pulsed condition, inset: natural logarithm of threshold current density against temperature in the ranges of 16–102°C. Reproduce from [98].

temperature T_0 of 32 K between 16–102°C. This result is further improved by K. Li et al. with an optimized DFLs and QDs [99].

As shown in **Figure 26a**, four repeats of $In_{0.18}Ga_{0.82}As$/GaAs SLSs DFLs were well performed to annihilate TDs with total buffer thickness of ~2 μm. The active region of laser was consisted of five repeats of InAs/GaAs DWELL structure, realizing a room temperature peak PL emission of ~1308 nm with a linewidth of ~32 meV. A comparison of room temperature PL results of InAs/GaAs QD on

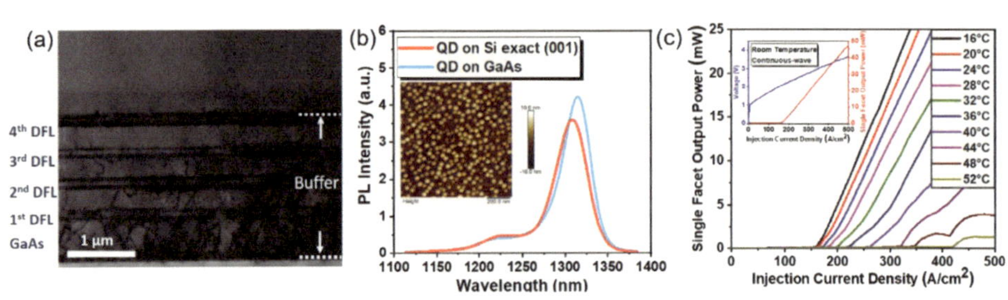

Figure 26.
(a) Cross-sectional TEM image for whole buffer; (b) A comparison of room temperature PL results, inset: an AFM image of uncapped InAs/GaAs QD layer; (c) Light-current characteristics of InAs/GaAs QD laser grown on Si exact (001) at various operation temperature, inset: light-current–voltage characteristic at room temperature. Reproduce from [99].

on-axis Si (001) and native GaAs substrates was shown in **Figure 26b**, the inset image shows an AFM image of uncapped InAs/GaAs QD layer with about $\sim 4 \times 10^{10}$ cm^{-2} dot density. The laser samples were fabricated into 50 μm × 3 mm broad-area laser devices. The characterization of laser devices was all measured under continuous wave. As shown in the inset image of **Figure 26c**, the threshold current density as low as ~160 A/cm^2 has been achieved at room temperature, which was improved compare to previous result. A single facet output power of 48 mW was obtained at an injection current density of 500 A/cm^2 without any thermal rollover. The threshold current density increased with the rising of operation temperature and laser operation was observed up to 52°C. The T_0 obtained was ~60.8 K between 16–36°C.

In order to investigate the defect tolerance of QD and QW structure, an InAs/GaAs QD laser directly grown on on-axis GaAs/Si (001) platform and an InGaAs QW laser in the same structure except active region were grown for comparison [100]. By further analyzing the performance of QD and QW laser and their thermal activation energy (E_a), the great characteristics of QD laser on dislocation tolerance and thermal reliability have been proved.

Temperature dependent PL measurements were performed for both QD and QW samples. As shown in **Figure 27a**, PL intensity of the QD sample at room-temperature was about six times lower than the PL intensity at 20 K. In contrast, the difference for the QW sample shown in **Figure 27b** was ~1000 times between 20 K and room temperature. Moreover, the integrated PL intensity for both samples was measured in order to estimate their E_a. The results were shown in **Figure 27c**, which were 240 meV and 35 meV for QD and QW lasers, respectively. The significantly higher E_a observed for the QD could contribute to its higher optical intensity at high temperatures. As shown in **Figure 28a, 25** μm × 3 mm broad-area lasers were fabricated for both QD and QW samples. The room-temperature characteristics of them under continuous-wave mode were illustrated in **Figure 28b**. The threshold current density of ~173 A/cm^2 for QD laser was achieved. In addition, over 100 mW single facet output power was obtained under injection current density of 670 A/cm^2. In contrast, there was no lasing observed for the QW device at room-temperature even at higher injection levels. After comparing with modelling results, this study indicated that QW laser cannot work properly above 10^7 cm^{-2} of TDD [100, 101]. **Figure 28c** presented a temperature dependent light-current curve of QD laser on Si (001). The highest continuous-wave operation was observed over 65°C. These results quantitively suggested that QD laser had its natural advantages on defect tolerance and temperature insensitivity. It also demonstrated that QD laser monolithically integrated on on-axis Si (001) substrate can be a promising on-chip optical source for Si photonics.

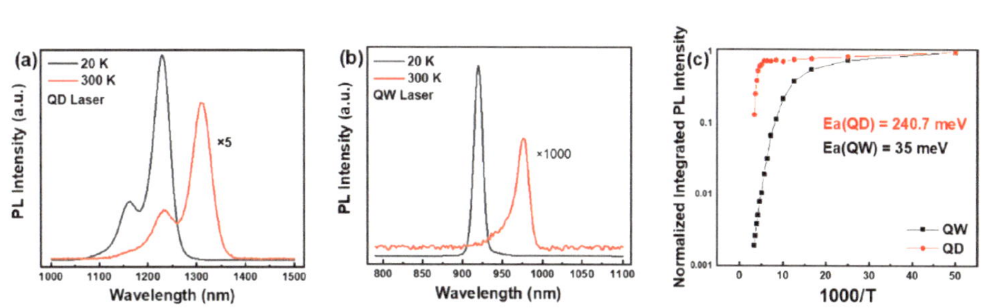

Figure 27.
Comparison of PL spectra at room-temperature (300 K) and 20 K for (a) the QD laser, and (b) the QW laser. (c) Temperature-dependent integrated PL intensities of the InAs QD and InGaAs QW lasers from the temperature region of 20 K to 300 K, showing Ea of both samples. Reproduce from [100].

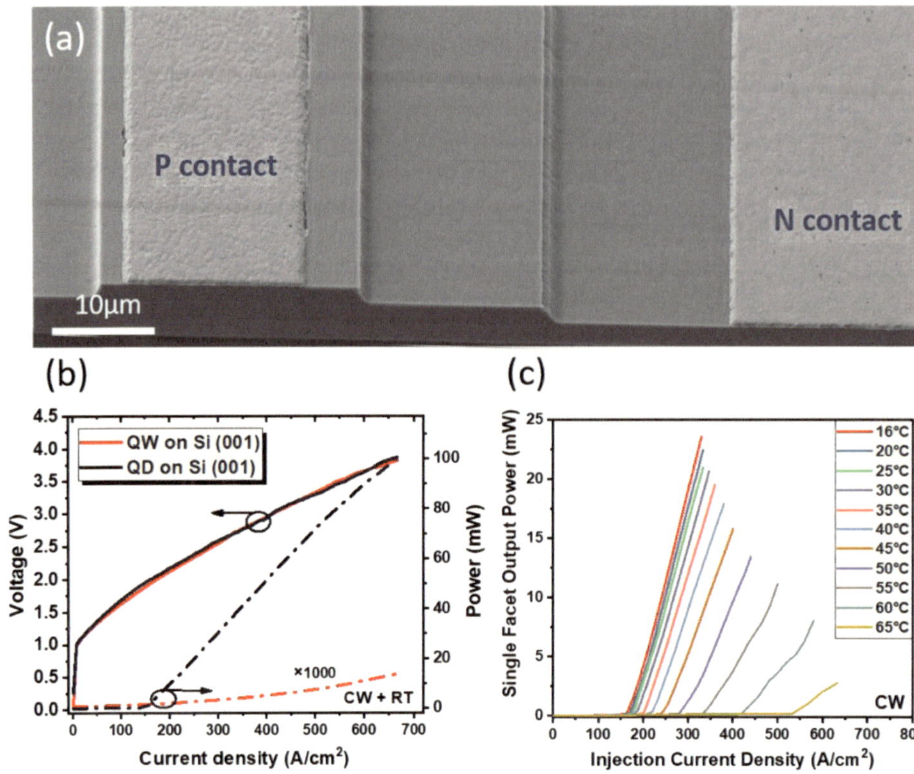

Figure 28.
(a) SEM image of an example of broad area laser fabricated by QD and QW samples with 25 μm ridge width and 3 mm cavity length. (b) Comparison of room-temperature light-current–voltage characteristics for QD and QW lasers directly grown on on-axis Si (001) substrate. (c) Temperature-dependent light-current measurement of the QD laser under continuous-wave mode. Reproduce from [100].

3.3 Microdisk QD laser grown on on-axis Si (001) substrate

Despite the outstanding progress has been made on edge-emitting QD lasers on on-axis GaAs/Si (001) substrate. For the dense integration with light source on Si that compatible to CMOS technique, microdisk lasers with small footprint has been considered as a promising approach for realizing nanophotonic integrated circuits. Additionally, compared with Fabry-Perot laser cavity, microdisk lasers also benefit from their advantages on low threshold and high quality factor which could bring less optical loss [106]. Recently, by applying the well-performed on-axis GaAs/Si (001) platform and optimized DFLs, a monolithically grown InAs/GaAs QD microdisk laser on on-axis Si (001) substrate with ultra-low threshold at room temperature was successfully demonstrated [107]. The device was optically pumped under continuous-wave mode. **Figure 29a** presented a schematic structure of this fabricated microdisk laser where the top of disk was the active region that consisted of three stacks of InAs/GaAs DWELL layers separated by 50 nm of GaAs space layer and 69 nm of AlGaAs cladding layer. A typical fabricated microdisk laser with disk diameter of 1.9 μm was shown in the SEM image of **Figure 29b**, which indicated a smooth etched surface with 73.5° sidewall tilt. The cross-sectional TEM image in **Figure 29c** shows the whole epilayer structure on on-axis GaAs/Si (001) substrate.

The collected lasing spectra for the microdisk laser with 1.9 μm diameter was illustrated in **Figure 30a**. The results presented a free spectral range of 76 nm –

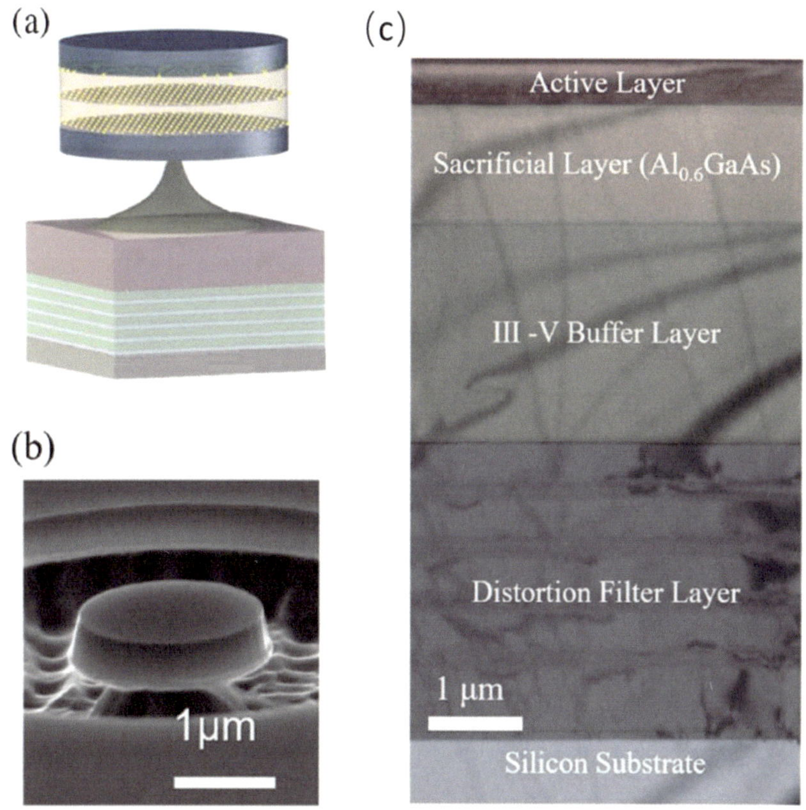

Figure 29.
(a) Schematic diagram of a QD microdisk laser fabricated on on-axis Si (001) substrate. (b) SEM image of a QD microdisk laser with 1.9 µm diameter. (c) Cross section TEM image of the epitaxial structure of QD microdisk laser on on-axis Si (001) substrate. Reproduce from [107].

89 nm between adjacent whispering gallery modes. Both ground state and excited state emission were observed. A main peak wavelength of 1263 nm was located at the first excited state. **Figure 30b** shows the collected intensity and linewidth as a function of input optical power for the corresponding peak emission at 1263 nm. An ultra-low threshold of 2.6 ± 0.4 µW and a clear narrowing trend of FWHM was obtained. The threshold of this result was even lower than the InAs QD microdisk lasers on native GaAs and InP substrates [109–111]. Additionally, the sample was fabricated into microdisk lasers with variable diameter from 1 µm to 2 µm. The corresponding threshold of main peak of microdisk lasers were presented as a function of diameter in **Figure 30c**. All the results of threshold were below 3.5 µW. The fluctuation of threshold versus the diameter of microdisk may result from the slight factor difference in fabrication process.

3.4 Continuous-wave QD photonic crystal lasers on on-axis Si (001)

As a promising ultra-compact on-chip light source, III-V photonic crystal lasers on Si benefits on their ultralow power consumption and small footprint. Most recently, Zhou et al. demonstrated an optically pumped InAs QD photonic crystal laser on on-axis GaAs/Si (001) substrate, which was the first monolithic integration of photonic crystal laser emitting at 1.3 µm on CMOS-compatible Si (001) substrate [108].
A single mode operation with ultra-low threshold down to $\sim$0.6 µm and a large

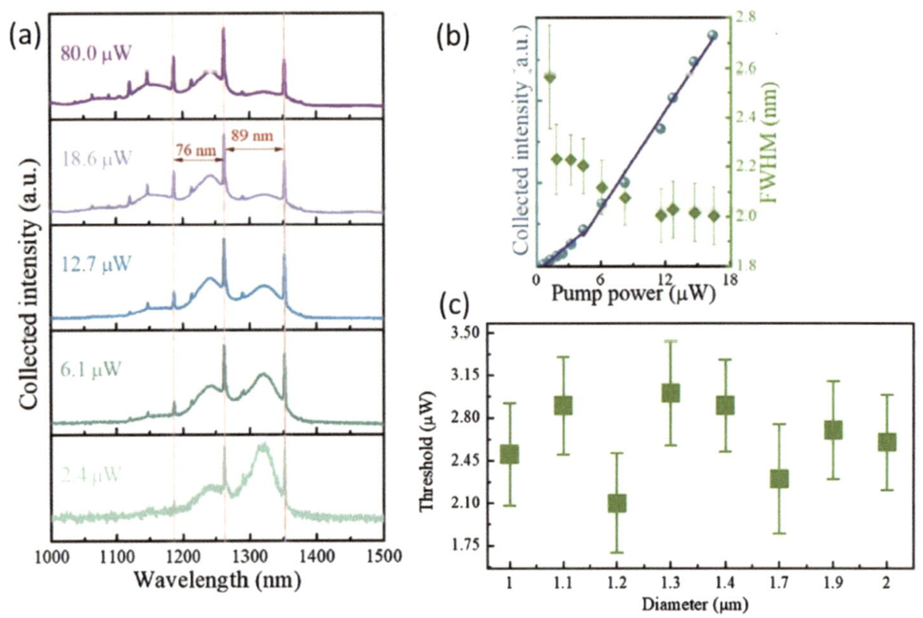

Figure 30.
(a) Collected intensity as a function of wavelength with different pumped power below and above the threshold of QD microdisk laser on on-axis Si (001). (b) The corresponding collected intensity and linewidth versus pumped power for the first excited state emission at 1263 nm. (c) Threshold of main lasing peak of QD microdisk laser as a function of various diameter. Reproduce from [108].

coupling efficiency for room temperature spontaneous emission under continuous-wave condition were achieved. 3D finite-difference time- domain (FDTD) simulation method was applied in order to obtain a high-quality factor for the resonance among QDs emission spectrum. **Figure 31a** shows a schematic structure of fabricated photonic crystal laser with 1 μm thickness of air slab underneath the cavity to enhance the vertical light confinement. The structure of active region that consists of four repeats of InAs/InGaAs/GaAs DWELL layers sandwiched by 50 nm GaAs space layers and 40 nm AlGaAs cladding layers was shown in **Figure 31b**. The collected intensity and linewidth of photonic crystal laser as a function of input power were shown in **Figure 31c**. The optically pumped QD photonic crystal lasers exhibited single-mode operation with an ultra-low threshold of ~0.6 μW. The inset image shows a peak wavelength at ~1306 nm with different pumped power. The Lorentzian fitting curve indicated a linewidth of ~0.68 nm and a calculated cavity quality factor of 2177. The soft turn on process shown in **Figure 31c** also presented a typical behavior of laser with high spontaneous emission coupling efficiency (β). The logarithmic plot of light–light curve with fitting results of this QD photonic crystal laser were shown in **Figure 31d**. It indicated the best fitting data obtained at β = 0.18, realizing a large spontaneous emission coupling efficiency under continuous-wave condition at room temperature.

3.5 Summary

QD laser on Si has attracted great research interests in recent years, which brings new approach for achieving efficient light source of Si-based photonics integration. These works discussed in this section with established epitaxy technique of APB-free on-axis GaAs/Si (001) platform, effective DFLs and optimized QD layers demonstrate that high-performance QD laser monolithically integrated on on-axis

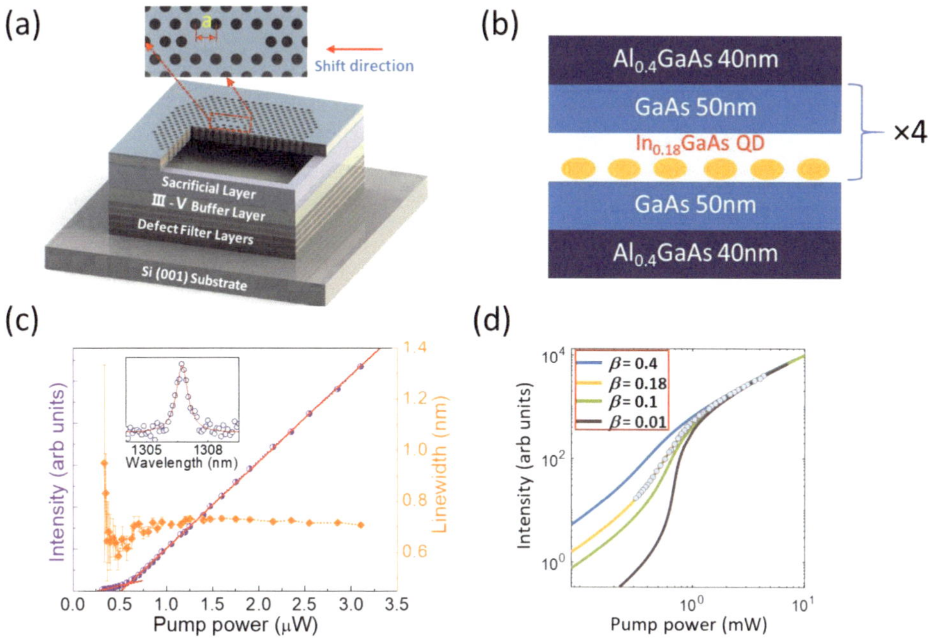

Figure 31.
(a) Schematic structure of QD photonic crystal laser on on-axis Si (001). (b) A diagram of active region in our photonic crystal laser. (c) Collected light–light curve and linewidth of the lasing peak at 1306 nm, inset: Lorentzian fitting of data below the threshold. (d) Logarithmic light–light plot of fitted and collected data. Reproduce from Ref. [108].

Si (001) substrate can be a promising on-chip optical source for Si photonics. Different approaches also provide new routes to form the basis of future monolithic light sources for the application of optical interconnects in large-scale silicon optoelectronics integrated circuits.

4. Conclusions

Heterogeneous integration of III–V compound semiconductors is promising to realize functionalities such as laser sources and photodetectors, and silicon based waveguides on Si platform. The direct heteroepitaxy of GaAs on nominal Si(100) wafers used by the microelectronics industry faces several issues to produce high quality material. In this chapter, we discussed the recent advances to tackle the formation of antiphase domains and to reduce the threading dislocation density. Currently, APB is no more an issue, as solutions have been proposed to obtain thin (<400 nm) GaAs film without APB, solutions based on dedicated cleaning and annealing processes of Si substrate before the GaAs epitaxy. The threading dislocations have hindered the development of GaAs devices on a Si CMOS platform and many solutions have been studied in th epast. We have reviewed the most efficient methods that used interchangeably the insertion of a Ge buffer between silicon and GaAs, the insertion of dislocation filter layers in the GaAs, or selective epitaxy in a cavity with a proper aspect ratio. All these progresses allowed reaching the range of 10^6–10^5 cm^{-2} TDD required to elaborate performant optoelectronics devices. Next we developed the fabrication of InAs QDs/GaAs laser emitters in the infrared region integrating GaAs buffer without APB grown on nominal Si(100) wafers and DFL to reduce the TDD. Different type of devices were fabricated such as broad area laser electrically pumped and operating at room temperature and up to 65°C, microdisk QDs lasers and continuous-wave QD photonic crystal lasers.

This paves the way towards the monolithic integration of optoelectronics and microelectronics functionalities on the same silicon CMOS platform, promising tremendous evolution in the data treatment and computing fields.

Author details

Mickael Martin[1], Thierry Baron[1]*, Yann Bogumulowicz[2], Huiwen Deng[3], Keshuang Li[3], Mingchu Tang[3] and Huiyun Liu[3]

1 University Grenoble Alpes, CNRS, CEA-LETI Minatec, Grenoble INP, LTM, Grenoble, France

2 University Grenoble Alpes, CEA-LETI, Grenoble, France

3 Department of Electronic and Electrical Engineering, University College London, Torrington Place, London, United Kingdom

*Address all correspondence to: thierry.baron@cea.fr

References

[1] Kunert B, Mols Y, Baryshniskova M, Waldron N, Schulze A, Langer R. How to Control Defect Formation in Monolithic III/V Hetero-Epitaxy on (100) Si? A Critical Review on Current Approaches. Semiconductor Science and Technology. **2018**;*33*(9):093002. DOI: https://doi.org/10.1088/1361-6641/aad655

[2] Vanderbilt D, Lee C. Energetics of Antiphase Boundaries in GaAs. Physical Review B. **1992**;*45*(19):11192-11201. DOI: https://doi.org/10.1103/PhysRevB.45.11192

[3] Rubel O, Baranovskii S. Formation Energies of Antiphase Boundaries in GaAs and GaP: An Ab Initio Study. International Journal of Molecular Sciences. **2009**;*10*(12):5104-5114. DOI: https://doi.org/10.3390/ijms10125104

[4] Akiyama M, Kawarada Y, Kaminishi K. Growth of GaAs on Si by MOVCD. Journal of Crystal Growth. **1984**;*68*(1): 21-26. DOI: https://doi.org/10.1016/0022-0248(84)90391-9

[5] Akiyama M, Kawarada Y, Ueda T, Nishi S, Kaminishi K. Growth of High Quality GaAs Layers on Si Substrates by MOCVD. Journal of Crystal Growth. **1986**;*77*(1–3):490-497. DOI: https://doi.org/10.1016/0022-0248(86)90342-8

[6] Xu HY, Guo YN, Wang Y, Zou J, Kang JH, Gao Q, et al. Effects of Annealing and Substrate Orientation on Epitaxial Growth of GaAs on Si. Journal of Applied Physics. **2009**;*106*(8): 083514. DOI: https://doi.org/10.1063/1.3248372

[7] Pehlke E, Tersoff J. Phase Diagram of Vicinal Si(001) Surfaces. Physical Review Letters. **1991**;*67*(10):1290-1293. DOI: https://doi.org/10.1103/PhysRevLett.67.1290

[8] Chadi DJ. Stabilities of Single-Layer and Bilayer Steps on Si(001) Surfaces. Physical Review Letters. **1987**;*59*(15): 1691-1694. DOI: https://doi.org/10.1103/PhysRevLett.59.1691

[9] Alerhand OL, Berker AN, Joannopoulos JD, Vanderbilt D, Hamers RJ, Demuth JE. Finite-Temperature Phase Diagram of Vicinal Si(100) Surfaces. Physical Review Letters. **1990**; *64*(20):2406-2409. DOI: https://doi.org/10.1103/PhysRevLett.64.2406

[10] Poon TW, Yip S, Ho PS, Abraham FF. Equilibrium Structures of Si(100) Stepped Surfaces. Physical Review Letters. **1990**;*65*(17):2161-2164. DOI: https://doi.org/10.1103/PhysRevLett.65.2161

[11] Martin M, Caliste D, Cipro R, Alcotte R, Moeyaert J, David S, et al. Toward the III–V/Si Co-Integration by Controlling the Biatomic Steps on Hydrogenated Si(001). Applied Physics Letters. **2016**;*109*(25):253103. DOI: https://doi.org/10.1063/1.4972394

[12] Zhang Z, Chen H, Bolding BC, Lagally MG. Vacancy Diffusion on Si (100)-(2×1). Physical Review Letters. **1993**;*71*(22):3677-3680. DOI: https://doi.org/10.1103/PhysRevLett.71.3677

[13] Kim E, Chen C, Pang T, Lee YH. Ordering of Dimer Vacancies on the Si (100) Surface. Physical Review B. **1999**; *60*(12):8680-8685. DOI: https://doi.org/10.1103/PhysRevB.60.8680

[14] Tavernier A, Favennec L, Chevolleau T, Jousseaume V. Innovative Gap-Fill Strategy for 28 Nm Shallow Trench Isolation. ECS Transactions. **2012**;*45*(3):225-232. DOI: https://doi.org/10.1149/1.3700888

[15] Kunert B, Németh I, Reinhard S, Volz K, Stolz W. Si (001) Surface Preparation for the Antiphase Domain Free Heteroepitaxial Growth of GaP on Si Substrate. Thin Solid Films. **2008**;*517*

(1):140-143. DOI: https://doi.org/10.1016/j.tsf.2008.08.077

[16] Brückner S, Döscher H, Kleinschmidt P, Supplie O, Dobrich A, Hannappel T. Anomalous Double-Layer Step Formation on Si(100) in Hydrogen Process Ambient. Physical Review B. **2012**;*86*(19):195310. DOI: https://doi.org/10.1103/PhysRevB.86.195310

[17] Alcotte R, Martin M, Moeyaert J, Cipro R, David S, Bassani F, et al. Epitaxial Growth of Antiphase Boundary Free GaAs Layer on 300 Mm Si(001) Substrate by Metalorganic Chemical Vapour Deposition with High Mobility. APL Materials. **2016**;*4*(4):046101. DOI: https://doi.org/10.1063/1.4945586

[18] Uen W-Y, Li Z-Y, Huang Y-C, Chen M-C, Yang T-N, Lan S-M, et al. Heteroepitaxial Growth of GaAs on Si by MOVPE Using A-GaAs/a-Si Double-Buffer Layers. Journal of Crystal Growth. **2006**;*295*(2):103-107. DOI: https://doi.org/10.1016/j.jcrysgro.2006.07.026

[19] Zhou X, Pan J, Liang R, Wang J, Wang W. Epitaxy of GaAs Thin Film with Low Defect Density and Smooth Surface on Si Substrate. Journal of Semiconductors. **2014**;*35*(7):073002. DOI: https://doi.org/10.1088/1674-4926/35/7/073002

[20] Yu HW, Chang EY, Yamamoto Y, Tillack B, Wang WC, Kuo CI, et al. Effect of Graded-Temperature Arsenic Prelayer on Quality of GaAs on Ge/Si Substrates by Metalorganic Vapor Phase Epitaxy. Applied Physics Letters. **2011**;*99*(17):171908. DOI: https://doi.org/10.1063/1.3656737

[21] Lin AC, Fejer MM, Harris JS. Antiphase Domain Annihilation during Growth of GaP on Si by Molecular Beam Epitaxy. Journal of Crystal Growth. **2013**;*363*:258-263. DOI: https://doi.org/10.1016/j.jcrysgro.2012.10.055

[22] Cornet C, Charbonnier S, Lucci I, Chen L, Létoublon A, Alvarez A, et al. Zinc-Blende Group III-V/Group IV Epitaxy: Importance of the Miscut. Phys. Rev. Mater. **2020**;*4*(5):053401. DOI: https://doi.org/10.1103/PhysRevMaterials.4.053401

[23] Ohta K, Kojima T, Nakagawa T. Anisotropic Surface Migration of Ga Atoms on GaAs (001). Journal of Crystal Growth. **1989**;*95*(1–4):71-74. DOI: https://doi.org/10.1016/0022-0248(89)90354-0

[24] Horikoshi, Y. Migration-Enhanced Epitaxy of GaAs and AIGaAs. 21.

[25] Shitara, T.; Neave, J. H.; Joyce, B. A. As/Ga Ratio Dependence of Ga Adatom Incorporation Kinetics at Steps on Vicinal GaAs(OO1) Surfaces. 5.

[26] Bogumilowicz Y, Hartmann JM, Cipro R, Alcotte R, Martin M, Bassani F, et al. Anti-Phase Boundaries–Free GaAs Epilayers on "Quasi-Nominal" Ge-Buffered Silicon Substrates. Applied Physics Letters. **2015**;*107*(21):212105. DOI: https://doi.org/10.1063/1.4935943

[27] Tang M, Park J-S, Wang Z, Chen S, Jurczak P, Seeds A, et al. Integration of III-V Lasers on Si for Si Photonics. Progress in Quantum Electronics. **2019**;*66*:1-18. DOI: https://doi.org/10.1016/j.pquantelec.2019.05.002

[28] George I, Becagli F, Liu HY, Wu J, Tang M, Beanland R. Dislocation Filters in GaAs on Si. Semiconductor Science and Technology. **2015**;*30*(11):114004. DOI: https://doi.org/10.1088/0268-1242/30/11/114004

[29] Chen S, Li W, Wu J, Jiang Q, Tang M, Shutts S, et al. Electrically Pumped Continuous-Wave III–V Quantum Dot Lasers on Silicon. Nature Photonics. **2016**;*10*(5):307-311. DOI: https://doi.org/10.1038/nphoton.2016.21

[30] Lasers QD, Ustinov V. In: publications O s, editor. M. Oxford: Oxford University Press; 2003

[31] Selvidge, J.; Norman, J.; Hughes, E. T.; Shang, C.; Jung, D.; Taylor, A. A.; Herrick, R.; Bowers, J. E.; Mukherjee, K. Defect Filtering for Thermal Expansion Induced Dislocations in III-V Lasers on Silicon. 18.

[32] Tang M, Chen S, Wu J, Jiang Q, Kennedy K, Jurczak P, et al. Optimizations of Defect Filter Layers for 1.3-Mm InAs/GaAs Quantum-Dot Lasers Monolithically Grown on Si Substrates. IEEE Journal of Selected Topics in Quantum Electronics. **2016**;*22*(6):50-56. DOI: https://doi.org/10.1109/JSTQE.2016.2551941

[33] Yamaguchi M. Dislocation Density Reduction in Heteroepitaxial III-V Compound Films on Si Substrates for Optical Devices. Journal of Materials Research. **1991**;*6*(2):376-384. DOI: https://doi.org/10.1557/JMR.1991.0376

[34] Tachikawa M, Yamaguchi M. Film Thickness Dependence of Dislocation Density Reduction in GaAs-on-Si Substrates. Applied Physics Letters. **1990**;*56*(5):484-486. DOI: https://doi.org/10.1063/1.102773

[35] Wang, G.; Loo, R.; Simoen, E.; Souriau, L.; Caymax, M.; Heyns, M. M.; Blanpain, B. A Model of Threading Dislocation Density in Strain-Relaxed Ge and GaAs Epitaxial Films on Si (100). *Appl Phys Lett* 4.

[36] Zogg, H. Dislocation Reduction by Glide in Epitaxial IV–VI Layers on Si Substrates. 5.

[37] Yang VK, Groenert M, Leitz CW, Pitera AJ, Currie MT, Fitzgerald EA. Crack Formation in GaAs Heteroepitaxial Films on Si and SiGe Virtual Substrates. Journal of Applied Physics. **2003**;*93*(7):3859-3865. DOI: https://doi.org/10.1063/1.1558963

[38] Beanland R, Dunstan DJ, Goodhew PJ. Plastic Relaxation and Relaxed Buffer Layers for Semiconductor Epitaxy. Advances in Physics. **1996**;*45*(2):87-146. DOI: https://doi.org/10.1080/00018739600101477

[39] Soga T, Hattori S, Sakai S, Takeyasu M, Umeno M. Characterization of Epitaxially Grown GaAs on Si Substrates with III-V Compounds Intermediate Layers by Metalorganic Chemical Vapor Deposition. Journal of Applied Physics. **1985**;*57*(10):4578-4582. DOI: https://doi.org/10.1063/1.335363

[40] Liu AY, Zhang C, Norman J, Snyder A, Lubyshev D, Fastenau JM, et al. High Performance Continuous Wave 1.3 μ m Quantum Dot Lasers on Silicon. Applied Physics Letters. **2014**;***104***(4):041104. DOI: https://doi.org/10.1063/1.4863223

[41] Tang M, Chen S, Wu J, Jiang Q, Dorogan VG, Benamara M, et al. 13-Mm InAs/GaAs Quantum-Dot Lasers Monolithically Grown on Si Substrates Using InAlAs/GaAs Dislocation Filter Layers. Optics Express. **2014**;*22*(10):11528. DOI: https://doi.org/10.1364/OE.22.011528

[42] Liu HY, Hopkinson M, Harrison CN, Steer MJ, Frith R, Sellers IR, et al. Optimizing the Growth of 1.3 Mm InAs/InGaAs Dots-in-a-Well Structure. Journal of Applied Physics. **2003**;*93*(5):2931-2936. DOI: https://doi.org/10.1063/1.1542914

[43] Liu HY, Sellers IR, Badcock TJ, Mowbray DJ, Skolnick MS, Groom KM, et al. Improved Performance of 1.3µm Multilayer InAs Quantum-Dot Lasers Using a High-Growth-Temperature GaAs Spacer Layer. Applied Physics Letters. **2004**;*85*(5):704-706. DOI: https://doi.org/10.1063/1.1776631

[44] Ward T, Sánchez AM, Tang M, Wu J, Liu H, Dunstan DJ, et al. Design Rules for Dislocation Filters. Journal of

Applied Physics. **2014**;***116***(6):063508. DOI: https://doi.org/10.1063/1.4892162

[45] Chen S, Tang M, Wu J, Jiang Q, Dorogan VG, Benamara M, et al. 1.3µm InAs/GaAs Quantum-Dot Laser Monolithically Grown on Si Substrates Using InAlAs/GaAs Dislocation Filter Layers. In: *2014 International Semiconductor Laser Conference*; IEEE: Palma de Mallorca, Spain. 2014. pp. 88-89. DOI: https://doi.org/10.1109/ISLC.2014.177

[46] Chen S, Tang M, Wu J, Jiang Q, Dorogan V, Benamara M, et al. Long-Wavelength InAs/GaAs Quantum-Dot Light Emitting Sources Monolithically Grown on Si Substrate. Photonics. **2015**; *2*(2):646-658. DOI: https://doi.org/10.3390/photonics2020646

[47] Tang M, Salamo G, Liu H, Dorogan VG, Jiang Q, Seeds AJ, et al. Optimisation of the Dislocation Filter Layers in 1.3-Mm InAs/GaAs Quantum-Dot Lasers Monolithically Grown on Si Substrates. IET Optoelectronics. **2015**;***9***(2):61-64. DOI: https://doi.org/10.1049/iet-opt.2014.0078

[48] Yang J, Bhattacharya P, Mi Z. High-Performance $\hbox{In}_{0.5}\hbox{Ga}_{0.5} \hbox{As/GaAs}$ Quantum-Dot Lasers on Silicon With Multiple-Layer Quantum-Dot Dislocation Filters. IEEE Transactions on Electron Devices. **2007**;***54***(11):2849-2855. DOI: https://doi.org/10.1109/TED.2007.906928

[49] El-Masry NA, Tarn JC, Karam NH. Interactions of Dislocations in GaAs Grown on Si Substrates with InGaAs-GaAsP Strained Layered Superlattices. Journal of Applied Physics. **1988**;***64***(7): 3672-3677. DOI: https://doi.org/10.1063/1.341409

[50] Wang J, Hu H-Y, Deng C, He Y-R, Wang Q, Duan X-F, et al. Defect Reduction in GaAs/Si Film with InAs Quantum-Dot Dislocation Filter Grown by Metalorganic Chemical Vapor Deposition. Chinese Physics B. **2015**;***24***(2):028101. DOI: https://doi.org/10.1088/1674-1056/24/2/028101

[51] Brunhes T, Boucaud P, Sauvage S, Aniel F, Lourtioz J-M, Hernandez C, et al. Electroluminescence of Ge/Si Self-Assembled Quantum Dots Grown by Chemical Vapor Deposition. Applied Physics Letters. **2000**;***77***(12):1822. DOI: https://doi.org/10.1063/1.1308526

[52] Wang Y, Wang B, Sasangka WA, Bao S, Zhang Y, Demir HV, et al. High-Performance AlGaInP Light-Emitting Diodes Integrated on Silicon through a Superior Quality Germanium-on-Insulator. Photonics Res. **2018**;***6***(4):290. DOI: https://doi.org/10.1364/PRJ.6.000290

[53] Kohen D, Bao S, Lee KH, Lee KEK, Tan CS, Yoon SF, et al. The Role of AsH3 Partial Pressure on Anti-Phase Boundary in GaAs-on-Ge Grown by MOCVD – Application to a 200mm GaAs Virtual Substrate. Journal of Crystal Growth. **2015**;***421***:58-65. DOI: https://doi.org/10.1016/j.jcrysgro.2015.04.003

[54] Hartmann JM, Abbadie A, Cherkashin N, Grampeix H, Clavelier L. Epitaxial Growth of Ge Thick Layers on Nominal and 6° off Si(0 0 1); Ge Surface Passivation by Si. Semiconductor Science and Technology. **2009**;***24***(5): 055002. DOI: https://doi.org/10.1088/0268-1242/24/5/055002

[55] Hartmann JM, Papon AM, Destefanis V, Billon T. Reduced Pressure Chemical Vapor Deposition of Ge Thick Layers on Si(001), Si(011) and Si(111). Journal of Crystal Growth. **2008**;***310***(24):5287-5296. DOI: https://doi.org/10.1016/j.jcrysgro.2008.08.062

[56] Bogumilowicz Y, Hartmann JM, Rochat N, Salaun A, Martin M, Bassani F, et al. Threading Dislocations in GaAs Epitaxial Layers on Various Thickness Ge Buffers on 300 Mm Si Substrates.

Journal of Crystal Growth. **2016**;***453***: 180-187. DOI: https://doi.org/10.1016/j.jcrysgro.2016.08.022

[57] Ayers JE. The Measurement of Threading Dislocation Densities in Semiconductor Crystals by X-Ray Diffraction. Journal of Crystal Growth. **1994**;***135***(1–2):71-77. DOI: https://doi.org/10.1016/0022-0248(94)90727-7

[58] Li Q, Ng KW, Lau KM. Growing Antiphase-Domain-Free GaAs Thin Films out of Highly Ordered Planar Nanowire Arrays on Exact (001) Silicon. Applied Physics Letters. **2015**; ***106***(7):072105. DOI: https://doi.org/10.1063/1.4913432

[59] Li Q, Lau KM. Epitaxial Growth of Highly Mismatched III-V Materials on (001) Silicon for Electronics and Optoelectronics. Progress in Crystal Growth and Characterization of Materials. **2017**;***63***(4):105-120. DOI: https://doi.org/10.1016/j.pcrysgrow.2017.10.001

[60] Krost A, Heinrichsdorff F, Schnabel F, Schatke K, Bimberg D, Cerva H. Optical and Crystallographic Properties of High Perfection InP Grown on Si (111). Journal of Electronic Materials. **1994**;***23***(2):135-139. DOI: https://doi.org/10.1007/BF02655259

[61] Paladugu M, Merckling C, Loo R, Richard O, Bender H, Dekoster J, et al. Site Selective Integration of III–V Materials on Si for Nanoscale Logic and Photonic Devices. Crystal Growth & Design. **2012**;***12***(10):4696-4702. DOI: https://doi.org/10.1021/cg300779v

[62] Baryshnikova M, Mols Y, Ishii Y, Alcotte R, Han H, Hantschel T, et al. Nano-Ridge Engineering of GaSb for the Integration of InAs/GaSb Heterostructures on 300 Mm (001) Si. Crystals. **2020**;***10***(4):330. DOI: https://doi.org/10.3390/cryst10040330

[63] Cipro R, Baron T, Martin M, Moeyaert J, David S, Gorbenko V, et al. Low Defect InGaAs Quantum Well Selectively Grown by Metal Organic Chemical Vapor Deposition on Si(100) 300 Mm Wafers for next Generation Non Planar Devices. Applied Physics Letters. **2014**;***104***(26):262103. DOI: https://doi.org/10.1063/1.4886404

[64] David S, Roque J, Rochat N, Bernier N, Piot L, Alcotte R, et al. Spatially Correlated Structural and Optical Characterization of a Single InGaAs Quantum Well Fin Selectively Grown on Si by Microscopy and Cathodoluminescence Techniques. APL Materials. **2016**;***4***(5):056102. DOI: https://doi.org/10.1063/1.4949761

[65] Rouviere J-L, Béché A, Martin Y, Denneulin T, Cooper D. Improved Strain Precision with High Spatial Resolution Using Nanobeam Precession Electron Diffraction. Applied Physics Letters. **2013**;***103***(24):241913. DOI: https://doi.org/10.1063/1.4829154

[66] Vigouroux MP, Delaye V, Bernier N, Cipro R, Lafond D, Audoit G, et al. Strain Mapping at the Nanoscale Using Precession Electron Diffraction in Transmission Electron Microscope with off Axis Camera. Applied Physics Letters. **2014**;***105***(19):191906. DOI: https://doi.org/10.1063/1.4901435

[67] Wang Z, Junesand C, Metaferia W, Hu C, Wosinski L, Lourdudoss S. III–Vs on Si for Photonic Applications—A Monolithic Approach. Materials Science and Engineering B. **2012**;***177***(17):1551-1557. DOI: https://doi.org/10.1016/j.mseb.2011.12.006

[68] Czornomaz L, Uccelli E, Sousa M, Deshpande V, Djara V, Caimi D, et al. Confined Epitaxial Lateral Overgrowth (CELO): A Novel Concept for Scalable Integration of CMOS-Compatible InGaAs-on-Insulator MOSFETs on Large-Area Si Substrates. *Symposium on VLSI Technology (VLSI Technology)*; IEEE: Kyoto, Japan, 2015. *2015*:T172-

T173. DOI: https://doi.org/10.1109/VLSIT.2015.7223666

[69] *Future Directions in Silicon Photonics*; ELSEVIER ACADEMIC PRESS: S.l., 2019; Vol. 101.

[70] Streshinsky M, Ding R, Liu Y, Novack A, Yang Y, Ma Y, et al. Low Power 50 Gb/s Silicon Traveling Wave Mach-Zehnder Modulator near 1300 Nm. Optics Express. **2013**;*21*(25):30350. DOI: https://doi.org/10.1364/OE.21.030350

[71] Yin T, Cohen R, Morse MM, Sarid G, Chetrit Y, Rubin D, et al. 31 GHz Ge N-i-p Waveguide Photodetectors on Silicon-on-Insulator Substrate. Optics Express. **2007**;*15*(21):13965. DOI: https://doi.org/10.1364/OE.15.013965

[72] Dai D, Bauters J, Bowers JE. Passive Technologies for Future Large-Scale Photonic Integrated Circuits on Silicon: Polarization Handling, Light Non-Reciprocity and Loss Reduction. Light Sci. Appl. **2012**;*1*(3):e1-e1. DOI: https://doi.org/10.1038/lsa.2012.1

[73] Michel J, Liu J, Kimerling LC. High-Performance Ge-on-Si Photodetectors. Nature Photonics. **2010**;*4*(8):527-534. DOI: https://doi.org/10.1038/nphoton.2010.157

[74] Wu J, Jiang Q, Chen S, Tang M, Mazur YI, Maidaniuk Y, et al. Monolithically Integrated InAs/GaAs Quantum Dot Mid-Infrared Photodetectors on Silicon Substrates. ACS Photonics. **2016**;*3*(5):749-753. DOI: https://doi.org/10.1021/acsphotonics.6b00076

[75] Soref R. The Past, Present, and Future of Silicon Photonics. IEEE Journal of Selected Topics in Quantum Electronics. **2006**;*12*(6):1678-1687. DOI: https://doi.org/10.1109/JSTQE.2006.883151

[76] Liao M, Li W, Tang M, Li A, Chen S, Seeds A, et al. Selective Area Intermixing of III–V Quantum-Dot Lasers Grown on Silicon with Two Wavelength Lasing Emissions. Semiconductor Science and Technology. **2019**;*34*(8):085004. DOI: https://doi.org/10.1088/1361-6641/ab2c24

[77] Wu J, Chen S, Seeds A, Liu H. Quantum Dot Optoelectronic Devices: Lasers, Photodetectors and Solar Cells. J. Phys. Appl. Phys. **2015**;*48*(36):363001. DOI: https://doi.org/10.1088/0022-3727/48/36/363001

[78] Liao M, Chen S, Huo S, Chen S, Wu J, Tang M, et al. Monolithically Integrated Electrically Pumped Continuous-Wave III-V Quantum Dot Light Sources on Silicon. IEEE Journal of Selected Topics in Quantum Electronics. **2017**;*23*(6):1-10. DOI: https://doi.org/10.1109/JSTQE.2017.2693025

[79] Selvidge J, Norman J, Salmon ME, Hughes ET, Bowers JE, Herrick R, et al. Non-Radiative Recombination at Dislocations in InAs Quantum Dots Grown on Silicon. Applied Physics Letters. **2019**;*115*(13):131102. DOI: https://doi.org/10.1063/1.5113517

[80] Mukherjee K, Selvidge J, Jung D, Norman J, Taylor AA, Salmon M, et al. Recombination-Enhanced Dislocation Climb in InAs Quantum Dot Lasers on Silicon. Journal of Applied Physics. **2020**;*128*(2):025703. DOI: https://doi.org/10.1063/1.5143606

[81] Jung D, Norman J, Wan Y, Liu S, Herrick R, Selvidge J, et al. Recent Advances in InAs Quantum Dot Lasers Grown on On-Axis (001) Silicon by Molecular Beam Epitaxy. Physica Status Solidi A: Applications and Materials Science. **2019**;*216*(1):1800602. DOI: https://doi.org/10.1002/pssa.201800602

[82] Deppe DG, Holonyak N, Nam DW, Hsieh KC, Jackson GS, Matyi RJ, et al. Room-temperature Continuous Operation of p - n Al_x Ga_{1-x} As-GaAs Quantum Well Heterostructure Lasers

Grown on Si. Applied Physics Letters. **1987**;***51***(9):637-639. DOI: https://doi.org/10.1063/1.98371

[83] Deppe DG, Nam DW, Holonyak N, Hsieh KC, Matyi RJ, Shichijo H, et al. Stability of 300 K Continuous Operation of p - n $Al_x Ga_{1-x}$ As-GaAs Quantum Well Lasers Grown on Si. Applied Physics Letters. **1987**;***51***(16):1271-1273. DOI: https://doi.org/10.1063/1.98702

[84] Kaliski RW, Holonyak N, Hsieh KC, Nam DW, Lee JW, Shichijo H, et al. Continuous (300 K) Photopumped Laser Operation of $Al_x Ga_{1-x}$ As-GaAs Quantum Well Heterostructures Grown on Strained-layer GaAs on Si. Applied Physics Letters. **1987**;***50***(13):836-838. DOI: https://doi.org/10.1063/1.98006

[85] Hall DC, Deppe DG, Holonyak N, Matyi RJ, Shichijo H, Epler JE. Thermal Behavior and Stability of Room-temperature Continuous $Al_x Ga_{1-x}$ As-GaAs Quantum Well Heterostructure Lasers Grown on Si. Journal of Applied Physics. **1988**;***64***(6):2854-2860. DOI: https://doi.org/10.1063/1.341596

[86] Sugo M, Mori H, Sakai Y, Itoh Y. Stable Cw Operation at Room Temperature of a 1.5-μm Wavelength Multiple Quantum Well Laser on a Si Substrate. Applied Physics Letters. **1992**;***60***(4):472-473. DOI: https://doi.org/10.1063/1.106638

[87] Wu J, Tang M, Liu H. III–V Quantum Dot Lasers Epitaxially Grown on Si Substrates. In: *Nanoscale Semiconductor Lasers*. Elsevier; 2019. pp. 17-39. DOI: https://doi.org/10.1016/B978-0-12-814162-5.00002-9

[88] Gérard JM, Cabrol O, Sermage B. InAs Quantum Boxes: Highly Efficient Radiative Traps for Light Emitting Devices on Si. Applied Physics Letters. **1996**;***68***(22):3123-3125. DOI: https://doi.org/10.1063/1.115798

[89] Lacombe D, Ponchet A, Gérard J-M, Cabrol O. Structural Study of InAs Quantum Boxes Grown by Molecular Beam Epitaxy on a (001) GaAs-on-Si Substrate. Applied Physics Letters. **1997**;***70***(18):2398-2400. DOI: https://doi.org/10.1063/1.118863

[90] Kazi ZI, Egawa T, Umeno M, Jimbo T. Growth of InxGa1−xAs Quantum Dots by Metal–Organic Chemical Vapor Deposition on Si Substrates and in GaAs-Based Lasers. Journal of Applied Physics. **2001**;***90***(11):5463-5468. DOI: https://doi.org/10.1063/1.1375010

[91] Mi Z, Yang J, Bhattacharya P, Huffaker DL. Self-Organised Quantum Dots as Dislocation Filters: The Case of GaAs-Based Lasers on Silicon. Electronics Letters. **2006**;***42***(2):121. DOI: https://doi.org/10.1049/el:20063582

[92] Li L, Guimard D, Rajesh M, Arakawa Y. Growth of InAs/Sb:GaAs Quantum Dots on Silicon Substrate with High Density and Efficient Light Emission in the 1.3μm Band. Applied Physics Letters. **2008**;***92***(26):263105. DOI: https://doi.org/10.1063/1.2952594

[93] Wang T, Liu H, Lee A, Pozzi F, Seeds A. 13-Mm InAs/GaAs Quantum-Dot Lasers Monolithically Grown on Si Substrates. Optics Express. **2011**;***19***(12): 11381. DOI: https://doi.org/10.1364/OE.19.011381

[94] Lee AD, Qi J, Tang M, Zhang Y, Seeds AJ, Liu H. InAs/GaAs Quantum-Dot Lasers Monolithically Grown on Si, Ge, and Ge-on-Si Substrates. IEEE Journal of Selected Topics in Quantum Electronics. **2013**;***19***(4):1901107-1901107. DOI: https://doi.org/10.1109/JSTQE.2013.2247979

[95] Lee A, Jiang Q, Tang M, Seeds A, Liu H. Continuous-Wave InAs/GaAs Quantum-Dot Laser Diodes Monolithically Grown on Si Substrate with Low Threshold Current Densities. Optics Express. **2012**;***20***(20):22181.

DOI: https://doi.org/10.1364/OE.20.022181

[96] Wang Y, Chen S, Yu Y, Zhou L, Liu L, Yang C, et al. Monolithic Quantum-Dot Distributed Feedback Laser Array on Silicon. Optica. **2018**;*5*(5):528. DOI: https://doi.org/10.1364/OPTICA.5.000528

[97] Liao M, Chen S, Liu Z, Wang Y, Ponnampalam L, Zhou Z, et al. Low-Noise 13 Mm InAs/GaAs Quantum Dot Laser Monolithically Grown on Silicon. Photonics Res. **2018**;*6*(11):1062. DOI: https://doi.org/10.1364/PRJ.6.001062

[98] Chen S, Liao M, Tang M, Wu J, Martin M, Baron T, et al. Electrically Pumped Continuous-Wave 13 Mm InAs/GaAs Quantum Dot Lasers Monolithically Grown on on-Axis Si (001) Substrates. Optics Express. **2017**; *25*(5):4632. DOI: https://doi.org/10.1364/OE.25.004632

[99] Li K, Liu Z, Tang M, Liao M, Kim D, Deng H, et al. O-Band InAs/GaAs Quantum Dot Laser Monolithically Integrated on Exact (0 0 1) Si Substrate. Journal of Crystal Growth. **2019**;*511*:56-60. DOI: https://doi.org/10.1016/j.jcrysgro.2019.01.016

[100] Liu Z, Martin M, Baron T, Chen S, Seeds A, Penty R, et al. Origin of Defect Tolerance in InAs/GaAs Quantum Dot Lasers Grown on Silicon. J. Light. Technol. **2020**;*38*(2):240-248. DOI: https://doi.org/10.1109/JLT.2019.2925598

[101] Hantschmann C, Liu Z, Tang M, Chen S, Seeds AJ, Liu H, et al. Theoretical Study on the Effects of Dislocations in Monolithic III-V Lasers on Silicon. J. Light. Technol. **2020**;*38* (17):4801-4807. DOI: https://doi.org/10.1109/JLT.2020.2994300

[102] Liu AY, Peters J, Huang X, Jung D, Norman J, Lee ML, et al. Electrically Pumped Continuous-Wave 13 Mm Quantum-Dot Lasers Epitaxially Grown on on-Axis (001) GaP/Si. Optics Letters. **2017**;*42*(2):338. DOI: https://doi.org/10.1364/OL.42.000338

[103] Norman J, Kennedy MJ, Selvidge J, Li Q, Wan Y, Liu AY, et al. Electrically Pumped Continuous Wave Quantum Dot Lasers Epitaxially Grown on Patterned, on-Axis (001) Si. Optics Express. **2017**;*25*(4):3927. DOI: https://doi.org/10.1364/OE.25.003927

[104] Wan Y, Jung D, Norman J, Shang C, MacFarlane I, Li Q, et al. O-Band Electrically Injected Quantum Dot Micro-Ring Lasers on on-Axis (001) GaP/Si and V-Groove Si. Optics Express. **2017**;*25*(22):26853. DOI: https://doi.org/10.1364/OE.25.026853

[105] Wan Y, Norman J, Li Q, Kennedy MJ, Liang D, Zhang C, et al. 13 Mm Submilliamp Threshold Quantum Dot Micro-Lasers on Si. Optica. **2017**;*4*(8):940. DOI: https://doi.org/10.1364/OPTICA.4.000940

[106] Vahala, K. J. Optical Microcavities. *Nature 424* (6950), 839–846.

[107] Zhou T, Tang M, Xiang G, Fang X, Liu X, Xiang B, et al. Ultra-Low Threshold InAs/GaAs Quantum Dot Microdisk Lasers on Planar on-Axis Si (001) Substrates. Optica. **2019**;*6*(4):430. DOI: https://doi.org/10.1364/OPTICA.6.000430

[108] Zhou T, Tang M, Xiang G, Xiang B, Hark S, Martin M, et al. Continuous-Wave Quantum Dot Photonic Crystal Lasers Grown on on-Axis Si (001). Nature Communications. **2020**;*11*(1):977. DOI: https://doi.org/10.1038/s41467-020-14736-9

[109] Tian B, Wang Z, Pantouvaki M, Absil P, Van Campenhout J, Merckling C, et al. Room Temperature O-Band DFB Laser Array Directly Grown on (001) Silicon. Nano Letters. **2017**;*17*(1):

559-564. DOI: https://doi.org/10.1021/acs.nanolett.6b04690

[110] Cao, H.; Xu, J. Y.; Xiang, W. H.; Ma, Y. Optically Pumped InAs Quantum Dot Microdisk Lasers. 4.

[111] Zhu S, Shi B, Wan Y, Hu EL, Lau KM. 1.55 Mm Band Low-Threshold, Continuous-Wave Lasing from InAs/InAlGaAs Quantum Dot Microdisks. Optics Letters. **2017**;*42*(4):679-682. DOI: https://doi.org/10.1364/OL.42.000679

Chapter 9

IMPATT Diodes Based on GaAs for Millimeter Wave Applications with Reference to Si

Janmejaya Pradhan and Satya Ranjan Pattanaik

Abstract

The small signal characteristics of DDR IMPATTs based on GaAs designed to operate at mm-wave window frequencies such as 94, 140, and 220 GHz are presented in this chapter. Both the DC and Small signal performance of the above-mentioned devices are investigated by using a small signal simulation technique developed by the authors. The efficiency, output power and power density of GaAs IMPATT is higher than that of Si IMPATT. Results show that the DDR IMPATTs based on GaAs are most suitable for generation of RF power with maximum conversion efficiency up to 220 GHz. The noise behavior of GaAs IMPATT yield less noise as compared to Si IMPATT.

Keywords: GaAs, IMPATT diode, ionization rates, efficiency, noise

1. Introduction

The revolution of electronic device in 20th century is mostly based on silicon and is regarded as the first generation semiconductor. Before the beginning of 21st century gallium arsenide (GaAs) and indium phosphide (InP) have evolved as second generation semiconductors constituting the base for the wireless and information revolution. However, at the begin of the 21st century, silicon carbide (SiC) and gallium nitride (GaN) are emerge as the wide bandgap semiconductors can work at high temperature and at high voltage and they may be regarded as third generation semiconductors used in the electronic and optoelectronic industries. Moreover the superior properties of wide bandgap semiconductors and the recent rush of research on wide bandgap semiconductor based electronic devices; one might speculate that wide bandgap semiconductors like diamond, AlN, etc. may be the future generation semiconductors. However, all the semiconductors have their vital performance in the field of information and communication.

The tremendous growth in information and communication technology has resulted in demand for millions of channels simultaneously. In order to avoid the interference between individual communications, the frequency of operation has been increased to a high value (i.e. the microwave and millimeter wave range). The advancement in solid state devices has contributed significantly towards the feasibility of modern microwave and mm-wave systems. Among several solid state devices capable of producing millimeter wave, IMPATT (IMPact Avalanche Transit Time) diode is considered as a leading source of solid-state power. The high power generating capability and high efficiency of IMPATT diode makes it attractive both at commercial and military sectors. The device has the dominant characteristics over other microwave and millimeter wave sources both with respect to the frequency coverage and output power. Later the report of first experiment of the microwave oscillation [1] the efficiency and output power has been increased with frequency. And some of the records making output from IMPATT diode are 42 W of pulsed power at 96 GHz [2], 520 mW at 217 GHz [3] and a continuous wave (CW) power 980 mW near 100 GHz and 50 mW at 220 GHz [4] have been reported. Again IMPATT diode being fabricated from any semiconductor, it has made itself an attractive device for both theoretical as well as experimental study.

IMPATT diodes are well recognized two terminal solid-state devices to deliver sufficiently high power at both microwave and mm-wave frequency bands [5]. Silicon is the most popular base material for IMPATT diodes from the point of view of its advanced process technology [6–10]. However, GaAs is a vibrant base semiconductor for IMPATT diodes at the both microwave and mm-wave frequencies. Since early seventies, several researchers have fabricated IMPATT diodes based on GaAs and obtained higher DC to RF conversion efficiency and better avalanche noise performance of those as compared to their conventional Si counterparts [11–16].

This chapter looks at the benefits of GaAs in power electronics applications, reviews the current state of the art, and shows how it can be a strong and feasible candidate for IMPATT. It is also well known that at a given frequency the microwave and millimeter wave power output of an IMPATT diode is proportional to the square of the product of semiconductor critical field and carrier saturation velocity. Again heat generation and dissipation in IMPATT diodes can severely limit the performance of IMPATT diodes. GaAs is, therefore, an ideal semiconductor for IMPATT diodes over Si, because it offers higher (i) critical electric field, (ii) carrier saturation velocity and (iii) thermal conductivity. These properties can lead to high-performance IMPATT diodes for microwave and millimeter wave applications. Some theoretical work using drift–diffusion methods for IMPATT device simulation confirmed that GaAs devices operated in the pulsed mode can offer very high power in the short-wavelength part of the millimeter range. So in this chapter, we have explored the device properties of GaAs IMPATT diode using a small signal model for mm-wave applications around the design operating frequencies of 94, 140, 220 and 300 GHz. The power performance and noise behavior of the diode is determined and compared with the Si base double drift region (DDR) IMPATT diode.

2. Material parameter and design consideration of Si, GaAs IMPATT diodes

The material parameters take the vital role for the design of the diodes as well as the IMPATT operation. We have used the values of material parameters of the semiconductors under consideration i.e. the carrier ionization rate, saturation drift velocity of electron (v_{sn}), and hole (v_{sp}), mobility (μ), permittivity (ε_s) etc. obtained from the research reports [17–26]. The material parameters used for the computer simulation IMPATT diodes based on the semiconductors concerned are summarized in **Table 1**. Besides theses parameters for IMPATT it is required to consider about the diode area, junction temperature and the operating current density at the desired design frequency. In this case we have taken the uniform diode area and junction temperature, but the current density is taken as per the operating frequency.

Parameters	Si	GaAs
Energy band gap, E_g (eV)	1.12	1.42
Critical Electric Field, E_c ($\times 10^6$ V/m)	—	0.65
*A_n ($\times 10^8$ m^{-1})	0.62	5.6
*B_n ($\times 10^8$ V m^{-1})	1.31	2.41
* A_p ($\times 10^8$ m^{-1})	2.0	1.5
*B_p ($\times 10^8$ Vm^{-1})	2.17	1.57
*m	1	1
Saturation drift velocity of electron, v_{sn} ($\times$ 10^4 m/s)	10.5	10.0
Saturation drift velocity of holes, v_{sp} ($\times 10^4$ m/s)	8.1	10.0
Electron mobility, μ_n ($m^2V^{-1}S^{-1}$)	0.058	0.85
Hole mobility μ_p ($m^2V^{-1}S^{-1}$)	0.04	0.019
Permittivity, ε (10^{-11} F/m)	10.0	11.4

*$\alpha_n = A_n \exp[-(B_n/E)]^m$, $\alpha_p = A_p \exp[-(B_p/E)]^m$.

Table 1.
Material parameters of Si and GaAs semiconductors.

To design an IMPATT diode one needs to consider its efficiency, frequency of operation, low cost, low loss thermal and electrical constants, output power and it is also important to achieve the breakdown condition for IMPATT operation. Energy band gap, ionization rate, dielectric constant, thermal conductivity, saturation drift velocity of electron and hole and break down field are the key factors to acquire the simulation results of efficiency, breakdown voltage and the output power of IMPATT. Taking into account the suitability of all these material properties, we have used the design criteria as $W = 0.5\, v_{sn,sp}/f_d$; where W, $v_{sn,sp}$ and f_d are the total depletion layer width, saturation velocity of electrons and design operating frequency respectively and chosen the double drift region (DDR) optimized structure to explore the potential of GaAs IMPATT diode. The schematic diagram of the DDR structure is shown in **Figure 1**. This structure depends on the saturation velocity and design operating frequency.

The diodes have the doping distribution of the form n^+npp^+ and are designed to operate at a frequency of 94, 140 and 220 GHz. Each n and p-region has width as well as the total width for the active region which has been mentioned in the **Table 2**. The width of the n^+ and p^+ are negligible and are hence used for ohmic contacts. The doping concentration for each n and p region of all structures have been mentioned in **Table 2**, while the doping concentration for each n^+ and p^+ region are taken as 1.0×10^{26} m^{-3}. The optimized operating current density, junction temperature and diode area are taken for window frequency of 94 GHz, 140 GHz and 220 GHz and listed in **Table 2**. Again the junction temperature and diode area are taken as 300 K and 1.0×10^{-10} m^2 respectively.

Though the applications of IMPATT diode are mostly realized on the basis of double drift region structures, we have considered the symmetrical double drift region (DDR) IMPATT diode structures with doping distribution of the form n^+npp^+ as shown in **Table 2** for the DC, small signal and noise analysis. 1-D schematic diagram of the proposed DDR IMPATT diode structures are shown in **Figure 1**. The n^+ and p^+ regions of the diode are heavily doped with each having a doping concentration of 1.0×10^{26} m^{-3}. Each n- and p-regions has a moderate doping concentration for different materials based on the optimized current

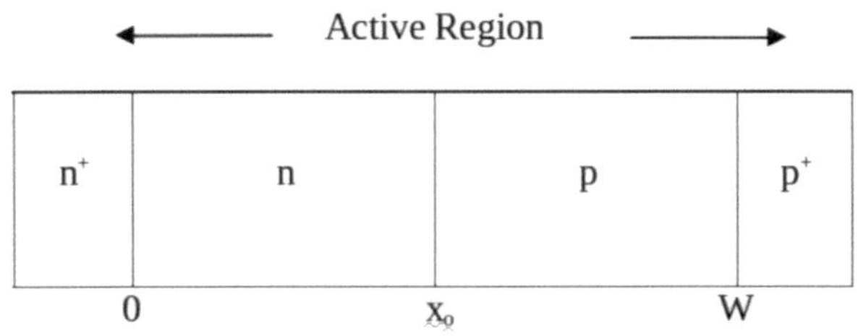

Figure 1.
A 1-D schematic diagram of the proposed DDR IMPATT diode.

Design frequency (GHz)	Materials	Width of active region (nm)		Doping concentrations ($\times10^{23}$ m^{-3})		Current density (J) ($\times10^8$ Am^{-2})
		n-region (W_n)	p-region (W_p)	n-region N_D	p-region N_A	
94	Si	555	430	0.80	0.85	3.2
	GaAs	530	530	0.65	0.65	2.0
140	Si	360	285	1.40	1.45	6.8
	GaAs	355	355	1.1	1.1	5.0
220	Si	245	185	2.70	2.75	15.0
	GaAs	225	225	2.00	2.00	7.3

Table 2.
Design parameters of Si and GaAs DDR IMPATT.

density as given in **Table 2**. The total active regions width is taken along with different space points of 1 nm each on both p-region and n-region. The values of doping concentrations and diode active region width are taken for optimum conversion efficiency and operation at atmospheric window frequencies of 94 GHz, 140 GHz and 220 GHz. The net doping concentration at any space point is hence determined by using the exponential and error function profiles.

3. Millimeter-wave properties of Si and GaAs DDR IMPATT diodes

The various properties of IMPATT diodes based on the fundamental semiconductor materials like Si and GaAs have been found by the simulation method. The properties like DC characteristics, small signal characteristics and noise behaviors have been computed in DDR structures based on Si and GaAs. The details of the results are discussed in the following sections.

3.1 DC characteristics

The computer simulation method [27] has been applied to a DDR structures IMPATT diode based on Si and GaAs and yields the results of different characteristics. The essential DC characteristics such as peak electric field (E_{max}), breakdown voltage (V_B), avalanche zone voltage (V_A), efficiency (η), avalanche zone width (X_A) and ratio of avalanche zone width to total depletion layer width (X_A/W) of the designed DDR IMPATTs are obtained from DC simulation.. The analysis of comparative description of the prospects of GaAs for IMPATT diode with reference to Si IMPATT diodes at different operating frequencies such as 94 GHz, 140 GHz

Design frequency f_d (GHz)	Material	Peak electric field E_{max} (×10^7 Vm^{-1})	Breakdown voltage V_B (volt)	Drift voltage V_D (volt)	X_A/W (%)	Efficiency η (%)
94	Si	4.90	23.35	8.72	38.80	11.89
	GaAs	4.81	29.48	12.28	41.42	13.26
140	Si	5.46	17.91	6.31	44.5	11.22
	GaAs	5.19	22.25	8.39	47.8	12.00
220	Si	6.27	13.18	4.29	45.48	10.36
	GaAs	5.83	14.91	5.44	48.22	11.62

Table 3.
DC properties of Si and GaAs DDR diodes at 94 GHz, 140 GHz and 220 GHz with design parameter of ***Table 2****.*

and 220 GHz are presented in **Table 3**. At different frequencies all the considered IMPATT diodes show different kinds of behavior.

The breakdown voltage for GaAs IMPATT diode shows the high value over Si IMPATT. This high value of break down voltage produces high RF power output as compared to Si IMPATT. Again, GaAs IMPATT provides more efficiency than Si IMPATT and the efficiency values are given in **Table 3**. The percentage of the ratio of avalanche zone width to total drift layer width (X_A/W) for all the diodes structure under consideration increases with higher operating frequencies. Higher value of X_A/W describes wider avalanche zone which leads to higher avalanche voltage (V_A) and lower drift zone voltage (V_D). In case of GaAs-based DDRs X_A/W is 42.36% at 94 GHz but it rises to 48.22% at 220 GHz which causes the decrease of efficiency (η) at 220 GHz. But in Si DDRs at 94 GHz, X_A/W is 42.7%, whereas it is 45.48% at 220 GHz and this leads to fall in efficiency (η) at higher mm-wave frequencies

3.2 Small signal characteristics

The DC output simulation parameters have been used as the input for simulation of small signal analysis. The significant high-frequency or small signal parameters obtained from this analysis are optimum frequency (f_p), peak negative conductance ($-G_o$), negative resistance (Z_R), RF power output (P_{RF}) and output power density (P_D). These parameters are obtained from the high-frequency simulation of GaAs and Si DDR IMPATTs at several biased current density and represented in **Table 4**.

Design frequency (GHz)	Materials	Negative conductance ($-G_o$) ×10^7 Sm^{-2}	Negative resistance ($-Z_R$) ×10^{-9} Ωm^2	Power density (P_D) ×10^9 Wm^{-2}
94	Si	2.55	17.3	1.74
	GaAs	1.91	8.49	2.07
140	Si	6.11	9.42	2.45
	GaAs	4.89	4.70	3.07
220	Si	15.6	3.96	3.40
	GaAs	10.4	1.59	3.92

Table 4.
Small signal characteristics of Si and GaAs DDR IMPATT diodes at design frequency of 94, 140 and 220 GHz.

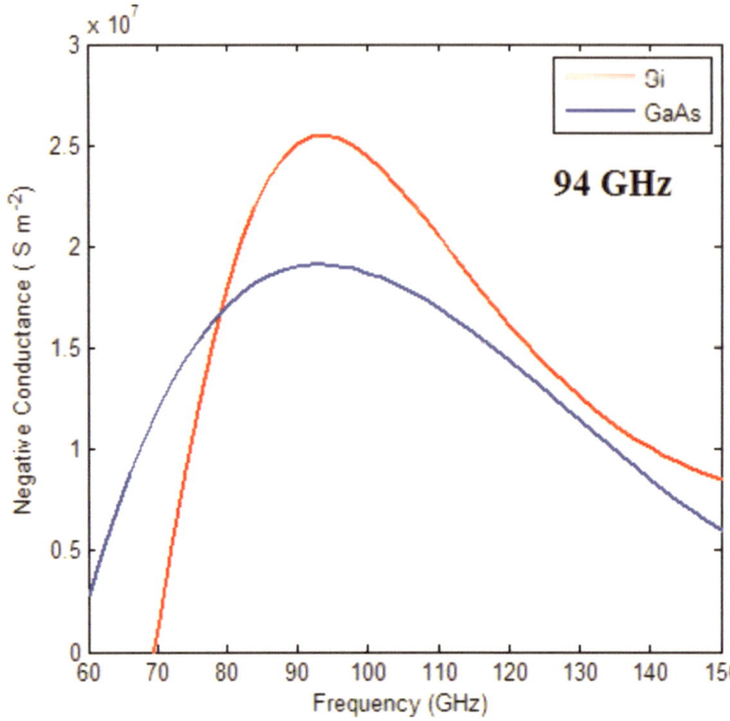

Figure 2.
Variation of negative conductance with frequency for DDR IMPATT operating at 94 GHz.

The diode negative conductance and negative resistance as a function of frequency for the different DDR IMPATT diodes of GaAs and Si are mentioned at different operating frequency. GaAs IMPPAT shows less negative conductance as compared to Si IMPATT. The diode negative conductance ($-G_o$) as a function of frequency for GaAs and Si DDR IMPATT is plotted in **Figures 2–4**. From the figures it is observed that, as the diodes are optimized with the current density, the peak of the negative conductance lies at the design operating frequencies 94 GHz, 140 GHz and 220 GHz and also it is noticed that the peak negative conductance of Si is remarkable higher. Subsequently in negative resistance case the behavior is directly reverse. The GaAs based IMPATT DDR diode gives less value of negative resistance ($-Z_R$) than that of Si DDR diodes.

The power density of GaAs bas IMPATT shows high value as compared to Si based IMPATT. The high value of power density indicates GaAs IMPATT diode is capable of high output power. Again it is noticed that the power density increases with increase in the operating frequency. This high power density of GaAs generates more noise in the device which is discussed in the subsequent section.

Over all, the variation of negative conductance with the different operating frequencies in **Figure 5** and it is also observed that the negative conductance of all the IMPATT diode based on Si and GaAs increases with the increase in operating frequency keeping area of the diode constant.

3.3 Noise properties

Since noise is an important aspect of IMPATT study, and hence, the noise characteristic of GaAs IMPATT diode has been analyzed and compared the results with Si based IMPATT diodes. In **Table 5**, the mean square noise voltage per band width of the three different diodes based on GaAs and Si at different

DOI: http://dx.doi.org/10.5772/intechopen.95874

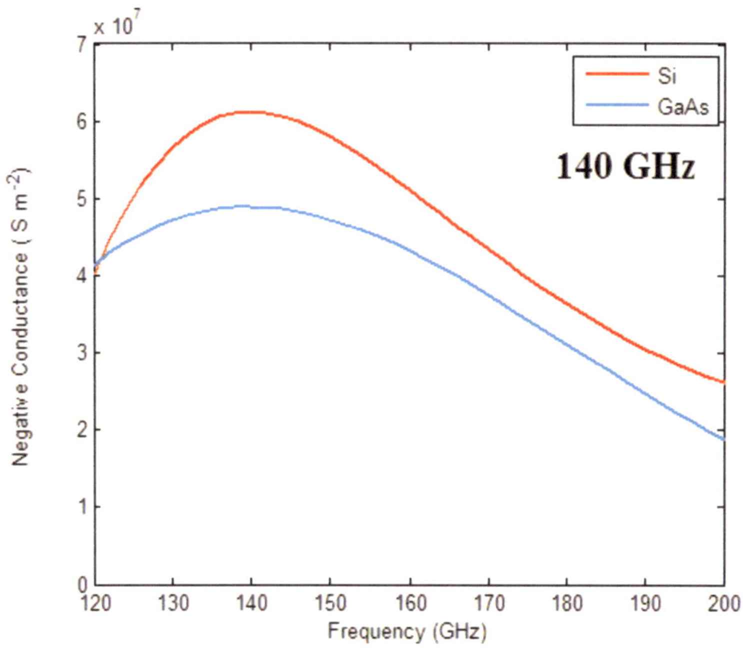

Figure 3.
Variation of negative conductance with frequency for DDR IMPATT operating at 140 GHz.

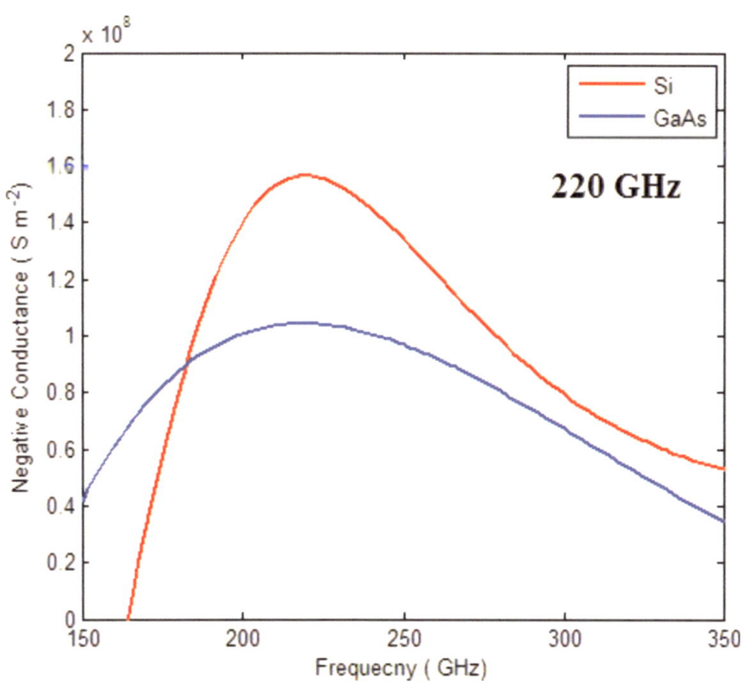

Figure 4.
Variation of negative conductance with frequency for DDR IMPATT operating at 220 GHz.

operating frequencies of 94 GHz, 140 GHz and 220 GHZ are represented. From **Table 5** it is seen that the peak values of mean square noise voltage per band width $(<v^2>/df)_{max}$ are found at the frequencies (f_p) of 75 GHz, 100 GHz and 160 GHz for Si for the operating frequency of 94 GHz, 140 GHz and 220 GHz,

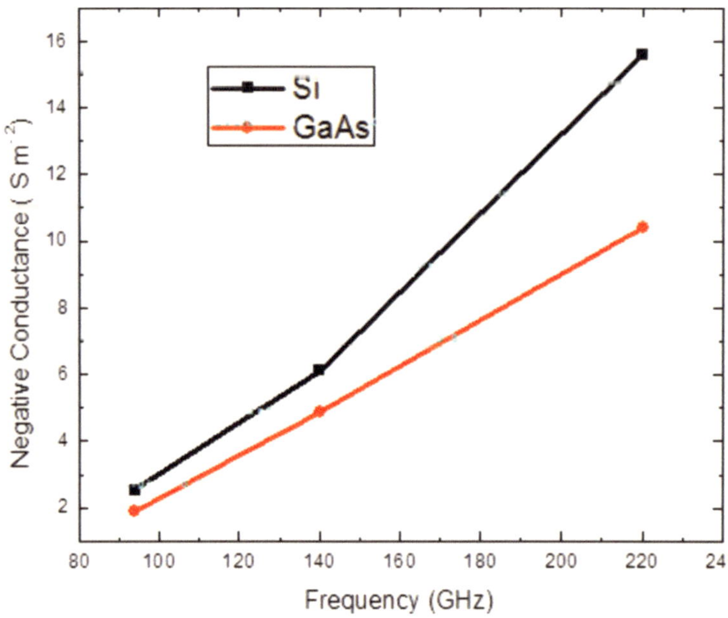

Figure 5.
Variation of negative conductance with operating frequency for DDR IMPATT diodes design to operate at 94, 140 and 220 GHz.

Material	Frequency at $(<v^2>/df)_{max}$ (GHz)	$(<v^2>/df)_{max}$ (V^2s)	f_d (GHz)	$(<v^2>/df)$ at f_d (V^2s)	NM at f_d (dB)
Si	75	1.59×10^{-14}	94	1.51×10^{-15}	27.23
	100	3.7×10^{-15}	140	6.51×10^{-16}	26.20
	160	8.95×10^{-16}	220	2.04×10^{-16}	24.94
GaAs	60	4.52×10^{-14}	94	5.54×10^{-16}	25.96
	110	9.75×10^{-15}	140	2.09×10^{-16}	24.28
	155	3.4×10^{-15}	220	5.98×10^{-17}	23.55

Table 5.
Noise properties of GaAs and Si DDR IMPATT diodes.

for GaAs IMPATT the frequencies of peak values of $<v^2>/df_{max}$ are 60 GHz, 110 GHz and 155 GHz. The corresponding values of mean square voltage per band width $(<v^2>/df)_{max}$ at the operating frequency of 94,140 and 220 GHz for all the diodes are given in **Table 5**. GaAs DDR IMPATT shows minimum mean square noise voltage per band width. The variation of mean square noise voltages per bandwidth (MSNVPBW) of the IMPATT diodes based on the semiconductors under consideration as a function of frequency are plotted in **Figures 6–8** for operating frequency of 94 GHz, 140 GHz and 220 GHz respectively.

Noise measure (NM), which is an indicator of noise to power ratio, is an important aspects for the study of noise behavior. The values of computed noise measure at the designed frequency are presented in **Table 5**. We have plotted noise measure as a function of frequency in **Figures 9–11** for both the IMPATT diodes based on GaAs and Si at different operating frequency of 94 GHz, 140 GHz and 220 GHz respectively. The comparative study of noise measure of Si and GaAs is

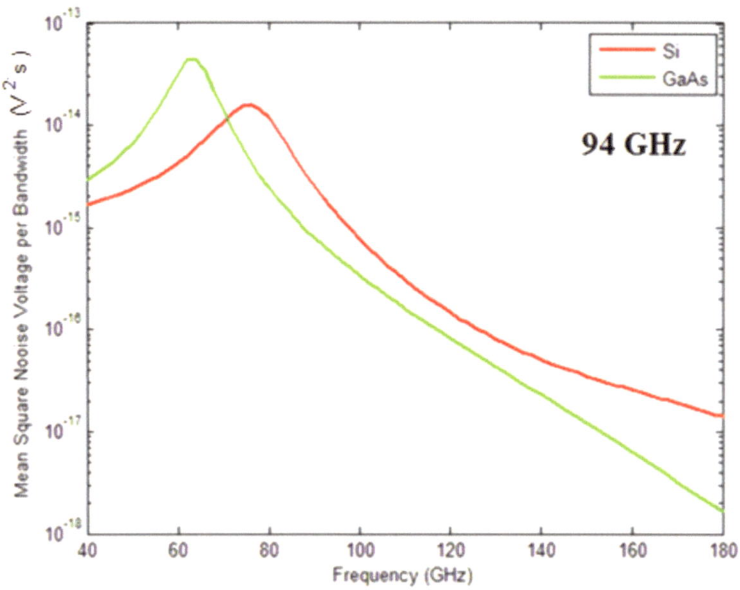

Figure 6.
Variation of mean square voltage per bandwidth with frequency for DDR IMPATT diodes operating at 94 GHz.

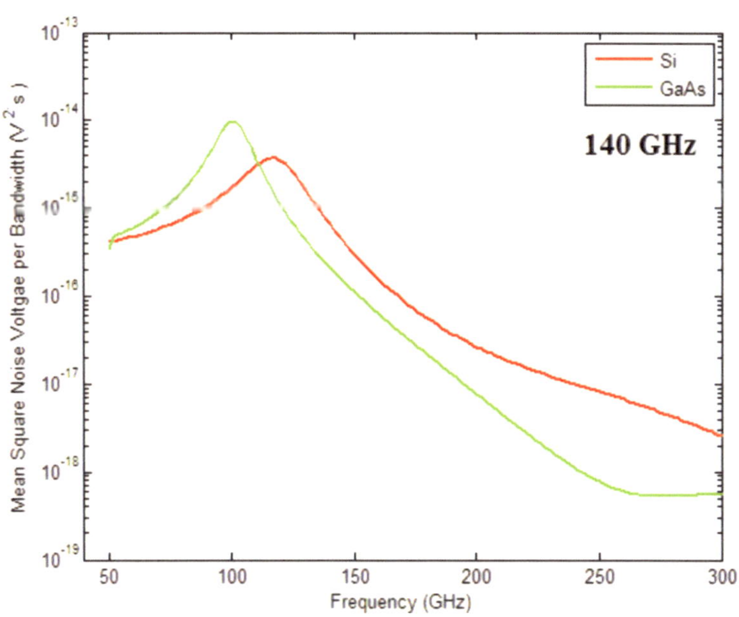

Figure 7.
Variation of mean square voltage per bandwidth with frequency for DDR IMPATT diodes operating at 140 GHz.

given in **Table 5**. GaAs IMPATT produces less noise as compare to Si IMPATT. The main reason for the low noise behavior of GaAs for a given electric field is the same value of the electron and hole ionization rates. Whereas, in Si the ionization rates are quite different. It may be mentioned here, the high power generation mechanism is such that if we wish to get more output power then we are supposed to get more noise.

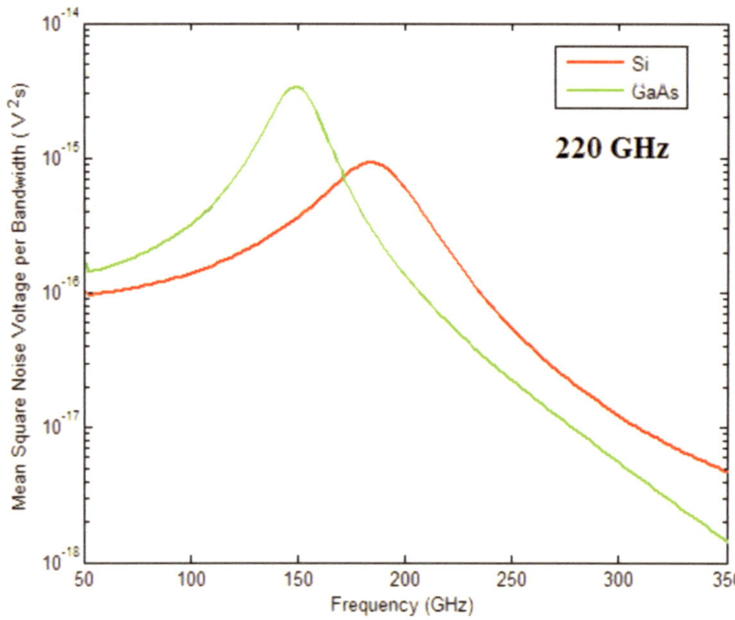

Figure 8.
Variation of mean square voltage per bandwidth with frequency for IMPATT DDR operating at 220 GHz.

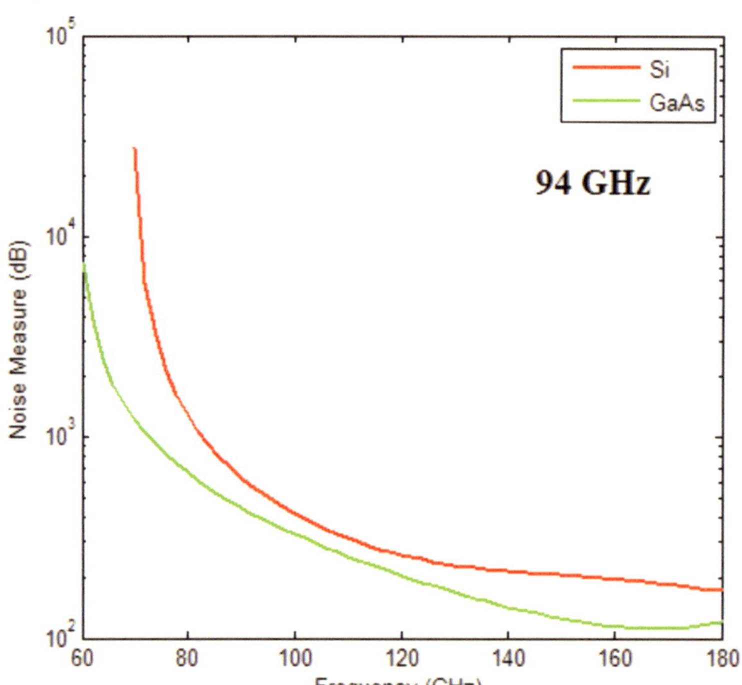

Figure 9.
Variation of noise measure with frequency for DDR IMPATT diodes operating at 94 GHz.

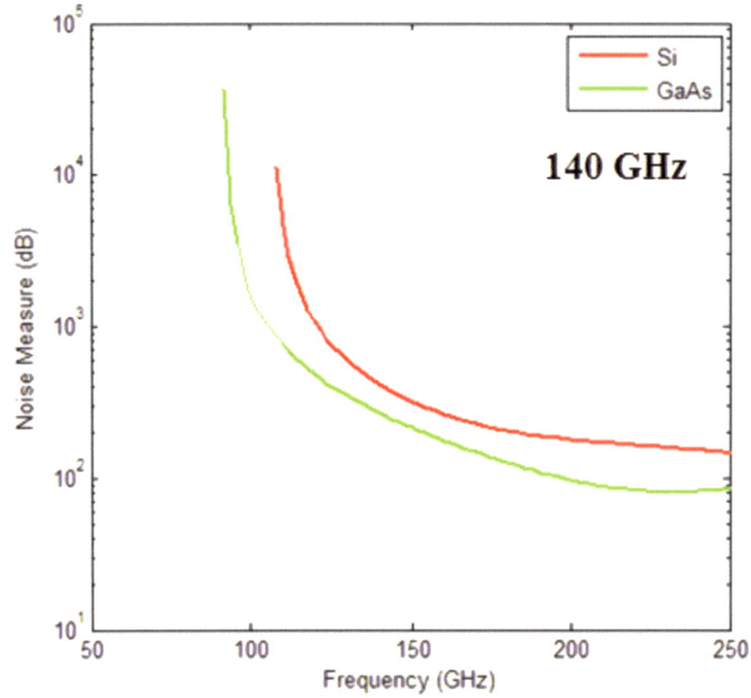

Figure 10.
Variation of noise measure with frequency for DDR IMPATT diodes operating at 140 GHz.

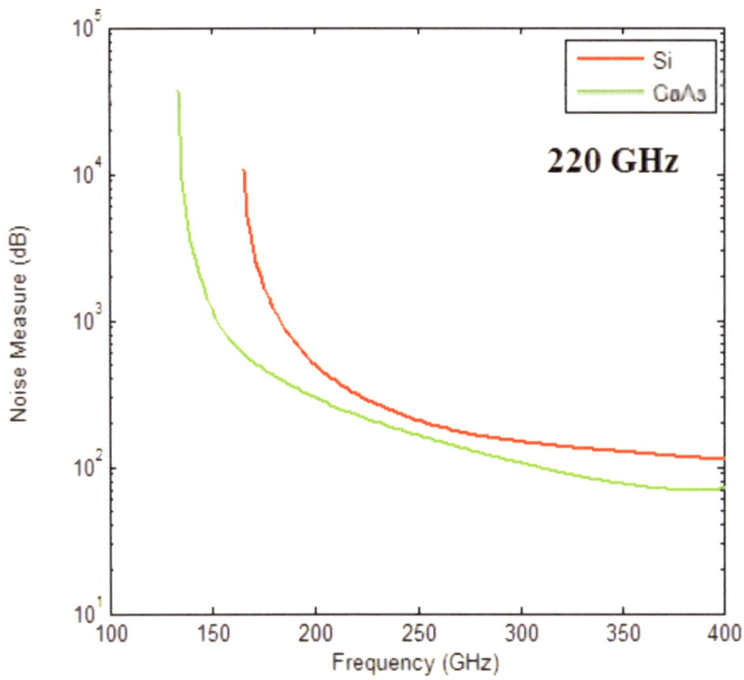

Figure 11.
Variation of noise measure with frequency for DDR IMPATT diodes operating at 220 GHz.

4. Conclusion

The small signal characteristics of DDR IMPATTs based on GaAs designed to operate at mm-wave window frequencies such as 94 GHz, 140 GHz, and 220 GHz are presented in this chapter. Both the DC and Small signal performance of the IMPATT diode based on GaAs are investigated by simulation technique. The results show that the DDR IMPATTs based on GaAs are most suitable device for generation of RF power with maximum conversion efficiency up to 220 GHz. However, at higher mm-wave frequencies, the DDR IMPATTs based on GaAs surpass the other semiconductor material for IMPATT not only for frequencies below 94 GHz but also above 94 GHz. Thus, GaAs is a vibrant base semiconductor for IMPATT diodes at the both microwave and mm-wave frequencies and also for higher mm-wave frequencies greater than 94 GHz. It is worthy indication of the future of communication technology.

Author details

Janmejaya Pradhan[1]* and Satya Ranjan Pattanaik[2]

1 Department of Physics, Rabindra Vidya Niketan, Keonjhar, India

2 National Institute of Science and Technology, Berhampur, Odisha, India

*Address all correspondence to: janmejaya74@gmail.com

References

[1] Johnston R L, Deloach B C and Cohen B G, "A Sillicon diode microwave oscillator", Bell Syst. Tech. J. 1965; 44;369

[2] Behr W and Luy J F, "High power operation mode of Pulsed Impatt diodes", IEEE Electron Device letters, 1990; 11, 206

[3] Fong T T and Kuno H J, "mm-wave pulsed Impatt sources", IEEE Trans. On MTT, 1979; 27,492-499

[4] Midford T A and Bernick R L, "mm-wave CW Impatt diodes and oscillators", IEEE Trasn. On MTT, 1979; 27,483-491

[5] Midford T A and Bernick R L, "Millimeter wave CW IMPATT diodes and oscillators," IEEE Transactions on Microwave Theory and Techniques, 1979; 27, (5), 483-492

[6] Luy J F, Casel A, Behr W, and Kasper E, "A 90-GHz double-drift IMPATT diode made with Si MBE," IEEE Transactions on Electron Devices, 1987; 34(5),1084-1089

[7] Dalle C, Rolland P A, and Lleti G, "Flat doping profile double-drift silicon IMPATT for reliable CW high power high-efficiency generation in the 94-GHz window," IEEE Transactions on Electron Devices, 1990; 37(1),227-236

[8] Luschas M, Judaschke R, and Luy J F, "Measurement results of packaged millimeter-wave silicon IMPATT diodes," in *Proceedings of the 27th International Conference on Infrared and Millimeter Waves, Conference Digest*, San Diego, Calif, USA, 2002; p. 135-136

[9] Luschas M, Judaschke R, and Luy J F., "Simulation and measurement results of 150 GHz integrated silicon IMPATT diodes," in *Proceedings of the IEEE MTT-S International Microwave Symposium Digest*, Wash, USA, June. 2002; 2, p.1269-1272

[10] Wollitzer M, Buechler J, Schäffler F, and Luy J F, "D-band Si-IMPATT diodes with 300 mW CW output power at 140 GHz," Electronics Letters, 1996; 32(2), 122-123

[11] Shih H D, Bayraktaroglu B, and Duncan W M, "Growth of millimeter-wave GaAs IMPATT structures by molecular beam epitaxy," Journal of Vacuum Science and Technology B: Microelectronics and Nanometer Structures, 1982;1(2),199-201

[12] Adlerstein M G and Chu S L G, "GaAs IMPATT diodes for 60 GHz," Electron Device Letters, 1984; 5(3), 97-98

[13] Eisele H, "Selective etching technology for 94 GHz GaAs IMPATT diodes on diamond heat sinks," Solid State Electronics, 1989;32(3),253-257

[14] Eisele H and Haddad G I, "GaAs TUNNETT diodes on diamond heat sinks for 100 GHz and above," IEEE Transactions on Microwave Theory and Techniques, 1995;43(1),210-213

[15] Tschernitz M and Freyer J, "140 GHz GaAs double-read IMPATT diodes," Electronics Letters, 1995; 31(7), 582-583

[16] Pradhan J, Swain S K, Pattanaik S R and Dash G N, "Potential of GaP as millimeter wave IMPATT diode with reference to Si, GaAs and GaN", Journal of Infrared and Millimeter wave, 2019;.34(4)

[17] Kyuregyan A S and Yurkov S N, "Room-temperature avalanche breakdown voltages of Si, Ge, SiC, GaAs, GaP and InP,"Sov. Phys. Semicond. 1989; 23(10), 1126-1132

[18] Blakemore J S, "Semiconducting *and other major properties of gallium*

arsenide," J. Appl. Phys. 1982; 53(10), R123-R181

[19] Pozhela J and Reklaitis A, "*Electron* transport properties in GaAs at high electric fields," Solid State Electron. 1980;.23 (9), 927-933

[20] Dalal V L, Dreeben A B, and Triano A, "Temperature Dependence of Hole Velocity in p-GaAs", J. Appl. Phys. 1971; 42(7), 2864-2867

[21] Pearsall T P, Capasso F, Nahory R E, Pallack M A, and Chelikowsky J, "The band structure dependence of impact ionization by hot carriers in semiconductors, Solid State Electron," 1978; 21,297-302

[22] Maes W, De Meyer K, and Van Overstraeten R, "Impact ionization in silicon," Solid State Electron., 1990; 33(6),705-718

[23] Grant W N, "*Electron* and Hole ionization rates in epitaxial silicon at high electric fields", Solid State Electron., 1973; 16(10),1189-1203

[24] Jacoboni C, Canali C, Ottaviani G and Quaranta A A, "A review of some charge transport properties of silicon", Solid State Electron., 1977; 20(2),77-89,.

[25] Arora V K, Mui D S L and Morkoc H, "High-field electron drift velocity and temperature in gallium phosphide", Appl. Phys, 1987;61,4703-4704

[26] Johnson R H and Eknoyan O, "High-field electron drift velocity measurements in gallium phosphide,"J. Appl. Phys. 1985;58(3), 1402-1403

[27] Dash G N and Pati S P, "small signal computer simulation of IMPATT diodes including carrier diffusion", Semiconductor Sci. and Technol. (IOP Journal, UK) 1991; 6(5), 348-355